Dance, Ageing and Collaborative Arts-Based Research

Dance, Ageing and Collaborative Arts-Based Research contributes a critical and comprehensive perspective on the role of the arts – specifically dance – in enhancing the lives of older people.

The book focuses on the development of an innovative arts-based program for older adults and the collaborative process of exploring and understanding its impact in relation to ageing, social inclusion, and care. It offers a wide audience of readers a richer understanding of the role of the arts in ageing and life enrichment, critical contributions to theories of ageing and care, specific approaches to arts-based collaborative research, and an exploration of the impact of Sharing Dance from the perspective of older adults, artists, researchers, and community leaders.

Given the interdisciplinary and collaborative nature of this book, it will be of interest across health, social science, and humanities disciplines, including gerontology, sociology, psychology, geography, nursing, social work, and performing arts.

Rachel Herron is Associate Professor in the Department of Geography and Environment at Brandon University, Canada, and the Canada Research Chair in Rural and Remote Mental Health.

Rachel Bar is Director, Research and Health at Canada's National Ballet School.

Mark Skinner is Dean of Humanities and Social Sciences at Trent University, Canada, where he is also Professor of Geography.

Dance, Ageing and Collaborative Arts-Based Research

Edited by Rachel Herron, Rachel Bar, and Mark Skinner

Routledge
Taylor & Francis Group
LONDON AND NEW YORK

First published 2023
by Routledge
4 Park Square, Milton Park, Abingdon, Oxon OX14 4RN

and by Routledge
605 Third Avenue, New York, NY 10158

Routledge is an imprint of the Taylor & Francis Group, an informa business

British Library Cataloguing-in-Publication Data
A catalogue record for this book is available from the British Library

Library of Congress Cataloging-in-Publication Data
Names: Herron, Rachel, editor. | Bar, Rachel, editor. | Skinner, Mark William, 1975- editor.
Title: Dance, ageing and collaborative art-based research / Edited by Rachel Herron, Rachel Bar and Mark Skinner.
Description: Abingdon, Oxon ; New York, NY : Routledge, 2022. | Includes bibliographical references and index.
Identifiers: LCCN 2022030760 (print) | LCCN 2022030761 (ebook) | ISBN 9781032197555 (hardback) | ISBN 9781032197562 (paperback) | ISBN 9781003260691 (ebook)
Subjects: LCSH: Dance for older people--Physiological aspects. | Dance for older people--Social aspects. | Art therapy for older people. | Aging--Social aspects.
Classification: LCC GV1799.3 .D36 2022 (print) | LCC GV1799.3 (ebook) | DDC 792.8084/6--dc23/eng/20220923
LC record available at https://lccn.loc.gov/2022030760
LC ebook record available at https://lccn.loc.gov/2022030761

ISBN: 978-1-032-19755-5 (hbk)
ISBN: 978-1-032-19756-2 (pbk)
ISBN: 978-1-003-26069-1 (ebk)

DOI: 10.4324/9781003260691

Typeset in Times New Roman
by SPi Technologies India Pvt Ltd (Straive)

Contents

Illustrations

Figures

Tables

Foreword

"*Sharing Dance* meant getting together again"

Missy Drummond with Dennis Drummond

I am pleased to provide a foreword to this book that offers the perspective of an older adult who participated in the Sharing Dance Older Adults program. My husband Dennis and I got involved with Sharing Dance when the program was hosted in partnership with our local Alzheimer's Society. I had been trying to find a program, which I could go to with Dennis where he didn't feel like he was standing out and much different from others because of his Alzheimer's disease. When the program was advertised, I phoned and we were accepted to go, but we weren't sure just what was going to take place. Through the Sharing Dance Older Adult program, I met the editors and some of the contributors to this book and I learned the benefits of sharing dance for relaxation, laughter, and fun as well as socializing, developing confidence, and exercising mind and body.

Right off the bat, we found it very friendly and easy to want to go to. The staff at the active living centre looked after everybody and they just made you feel like you'd known them all along. The facilitator and instructors included everybody. Nobody was ever singled out. The program shown on the television was very informative, easy to follow, and not strenuous. Dennis and I found it a relaxing time with the other people.

In the past, Dennis was not a dancer. I loved going to dances and I just thoroughly love dancing, but the odd time I'd get Dennis up. Although Dennis didn't dance much in the past, he did love music. Music seems to be able to draw people together. It draws a lot of things out of people, and it took the shyness away from people trying to do some of the moves in public. In the program, there was toe-tapping and everything else! They all just seemed to take right into that, whether they were a dancer or not. The music really got them going and it was enjoyable to watch. The music took away from some fuss on what was "wrong with" Dennis. Dennis was learning too, again, using all his arms and legs while having a good time with it.

Dennis opened up a lot and talked. When he was at the program, he got to discuss previous jobs that a lot of them had or if they were farmers, he'd talk about the farms with them and how things used to be back in the 60s or 70s compared to what it is now. It did get his mind thinking back on things. Also, he looked forward to going and it kept him thinking, "We go tomorrow" he'd say, or "is it the day after tomorrow that we go?" so he knew he was going to go. He looked forward to that day every week. Driving home, he'd get a big chuckle out of somebody or something that was said, and he'd hash it over. We'd have a good belly laugh about something.

Sharing Dance made you use your feet, your elbows, your shoulders, things you never really thought of before to move, in a way that was enjoyable and a lot of fun. You found bones that moved that you didn't think you could ever move again and in a fun manner. Our overall experience with this program was that it gave us the confidence to try something. Once you get into your late 70s and 80s – you either think you shouldn't – or you can't – do those things. Some people think you're too old to even try. You're not. You're all in the same boat, you're all learning the same thing. Sharing Dance meant getting together again. It brought out the relaxed social side of so many people. Everybody had a smile on their face. Nobody had an embarrassed look, or I-can't-do-it look. They all looked like they were thoroughly enjoying it. I'd look over at somebody and I'd be trying to catch up in a step or something and I'd just shake my head and we'd start laughing. I learned to laugh at myself when I goofed. You make friendships and your whole self and body is benefiting from it, mentally and physically.

People went out of their way to make Sharing Dance possible for others to take part in. This book is a collaboration of those people. It shares the stories of people like Dennis and I but also the stories of the organizations, staff, and artists that worked together to make Sharing Dance possible and successful. It is my hope that by sharing the stories of individual participants and partner organizations, others will be inspired to develop and support more collaborative arts-based programs for older adults.

Preface

Now, more than ever, we need to reimagine how we enable the rights of older adults and enrich their lives. Older adults, especially those living with dementia, continue to experience infringements on their rights to meaningful participation in the social world. They face discriminatory policies, practices, and attitudes that require radical transformation. This has become painfully apparent in the wake of the COVID-19 pandemic, which has led to death, confinement, and segregation of many older adults internationally. Addressing unequal access to opportunities that promote older adults' health and broader well-being is a critical challenge in the 21st century. Collaborative arts-based programs and research will play an essential role in addressing this challenge.

The overarching goal of this book is to share the theoretical, methodological, and practical lessons learned from the *Improving Social Inclusion for People with Dementia and Carers through Sharing Dance* project. The book is the culmination of a four-year collaborative research project funded by the Canadian Institutes of Health Research (CIHR) and Alzheimer Society of Canada, which explored the development of an innovative Sharing Dance Older Adults program offered by Canada's National Ballet School (NBS). The specific objectives of the research project were to examine the experiences of older adults participating in the program, assess the effectiveness of the remote delivery of the program, and identify the challenges of expanding the program, all with the goal of understanding how this arts-based innovation could improve the social inclusion of older adults. Although the book focuses on a single project, the broader aim of the book is to advance collaborative arts-based ageing research and programming through sharing lessons learned from the project. We hope that the book will inspire everyone to take up Sharing Dance or other arts-based innovations to enrich the lives of older adults no matter what their abilities are or where they live.

A key message of this book is that the transformative potential of arts-based programs and ageing research cannot be achieved alone. Collaboration is critical to the success, sustainability, and adaptability of accessible arts-based programs as well as research that captures the complex processes and impacts of such programs. The research in this book was born out of a shared view that bringing together diverse approaches and perspectives would yield a more comprehensive understanding of the importance and impact of the arts for older adults and their communities as well as the challenges and limitations that stand in the way of arts-based innovations. Working with Canada's National Ballet School has been a great example of how transdisciplinary collaboration can address the complex needs of individuals, build community capacity, respond to societal demands, and offer examples to leaders in the arts, community and long-term care, and ageing research internationally. Commitment to what we have come to know as *creative collaboration* continues to propel forward the lessons learned from the research, their application, and further adaptation.

As editors, we are grateful to the team of partners, co-investigators, collaborators, and research assistants for their dedication, imagination, and adaptability throughout the *Improving Social Inclusion for People with Dementia and Carers through Sharing Dance* project and its culmination in this book. Foremost, we thank sincerely the older adults and carers from Peterborough, Ontario and Brandon, Manitoba for their openness to explore something new and share their experiences of Sharing Dance with us. We thank the Alzheimer Society of Peterborough, Kawartha Lakes, Northumberland & Haliburton for their advice and feedback in the early stages of the research and the Alzheimer Society of Manitoba for working closely with the research team throughout many adaptations of the Sharing Dance program. We are grateful for the support of Community Care Peterborough in hosting Sharing Dance in Ontario and Bayside Personal Care Home, Birch Lodge Personal Care Home, Carberry Plains Personal Care Home, Country Meadows Personal Care Home, Minnedosa 50 + Activity Centre, Minnedosa Personal Care Home, and Prairie Oasis Community Centre for hosting Sharing Dance in Manitoba. We thank these organizations and their leaders, staff, and volunteers for collaborating with us to make this research possible. We also are grateful for the team of research assistants and graduate students at Trent University, Brandon University and NBS who supported the project including Heidi Burns, Amber Colibaba, Sylvia Dick, Sophia Kim, An Kosurko, Justin Sutton, Stephan Warrener, and Meghan Wrathall.

Finally, it is important to acknowledge the excellent contributions of our collaborating authors, and the editorial support of Amber Colibaba and An Kosurko, without which this book would not have come to fruition. It is also important to acknowledge on behalf of Canada's National Ballet School the supporters of Sharing Dance Older Adults, especially the Public Health Agency of Canada, Joan and Jerry Lozinski through the Lozinski Centre for Community Dance, Michael and Karen Vukets Family Foundation, and The Jack Weinbaum Family Foundation. The publication of this book was supported by the Canada Research Chairs program, Public Health Agency of Canada, and Trent Centre for Aging & Society (Trent University).

Rachel Herron, Rachel Bar, Mark Skinner

Contributors

Ilkka Arminen is Professor of Sociology in the Faculty of Social Sciences at the University of Helsinki, Finland. He has worked as a director of a network on innovation research and is an Adjunct Professor of Technology Studies at Aalto University's Department of Design. He has conducted user-centred research on information and communication technologies, and science and technology studies from 3D environments to ubiquitous computing, from emergency communication to HRM solutions and from aviation information systems to domestic technologies, with a focus on the social and human aspects of new forms of media and technologies. He leads a Kone Foundation project (2017–2021) on ethnic relations and the distribution of interactional expertise and was an international collaborator on the Canadian Institutes of Health Research (CIHR)/Alzheimer Society of Canada *Improving Social Inclusion for People with Dementia and Carers through Sharing Dance* project.

Rachel Bar is Director, Research and Health at Canada's National Ballet School and Fellow at the Trent Centre for Aging & Society, Trent University, Canada. She is a graduate of Canada's National Ballet School's professional ballet program and danced professionally before attending university. She completed her Ph.D. in psychology as a Vanier Scholar at Toronto Metropolitan University (formerly Ryerson University), Canada. Her research explores the benefits of dance for older adult populations, dancer health and wellness, and the utility of arts-based knowledge translation of health research. She was Research Coordinator and Postdoctoral Fellow on the Canadian Institutes of Health Research (CIHR)/Alzheimer Society of Canada *Improving Social Inclusion for People with Dementia and Carers through Sharing Dance* project.

Cassy Borth is Director of Programs at peopleCare Oakcrossing Long-Term Care in London, Ontario. She has been with peopleCare since 2019 and brings with her close to a decade of recreation-related experience in the long-term care sector. Working with seniors has always been Cassy's passion and she enjoys coming up with new creative, impactful programming that makes a difference, adds value, and supports meaning and quality of life for residents. Cassy was a community partner on the Canadian

Institutes of Health Research (CIHR)/Alzheimer Society of Canada *Improving Social Inclusion for People with Dementia and Carers through Sharing Dance* project.

Sheena Campbell is Vice President, Communications and Engagement for peopleCare. She is an accomplished communications expert with over 20 years of experience leading communications, engagement, and marketing initiatives for long-term care, home and community care, and government. She provides strategic guidance and oversight to enable peopleCare's collaboration and sector leadership that contributes to a strong, integrated health system. She also elevates how to share and leverage family-owned peopleCare's operational excellence, values-based decision-making, and resident-centred focus to support their vision to change the world of senior living. She holds an Honours BA and completed with honours the Public Relations program at Toronto Metropolitan University (formerly Ryerson University). She is a long-time member of the International Association of Business Communicators (IABC) and a member of the Ontario Long-Term Care Association Advocacy Committee. Sheena was a community partner on the Canadian Institutes of Health Research (CIHR)/Alzheimer Society of Canada *Improving Social Inclusion for People with Dementia and Carers through Sharing Dance* project.

John Dalrymple is Executive Director of Canada's National Ballet School. He previously held several positions at the School, including Chief of External Affairs, Director of Strategic Initiatives, and Associate Director of Annual Giving, as well as Professor at Centennial College in the School of Hospitality, Tourism and Culture. From 2001 to 2010, he was on the management team of the Textile Museum of Canada. He holds degrees from York University and the University of Toronto and was a knowledge user collaborator on the Canadian Institutes of Health Research (CIHR)/ Alzheimer Society of Canada *Improving Social Inclusion for People with Dementia and Carers through Sharing Dance* project.

Missy Drummond and **Dennis Drummond** took part in the Sharing Dance Older Adults program in their community of Brandon, Manitoba. Missy is 83 and her husband Dennis is 85 years old. They have been married for over 60 years and they have enjoyed dancing together since they were in their teens. Missy is a partner in care to Dennis, who was diagnosed with dementia seven years ago. Although Dennis has moved from the community into a long-term facility, he still enjoys dancing and music.

Margaret (Margo) Dunphy was 72 when she took part in the Sharing Dancing Program. She is a retired Dietary Cook and she worked at the Minnedosa District Hospital for 25 years. She was the head cook for many of those years. She started going to the Sharing Dance program because her sister wanted to go and see what it was all about. Margo's sister had just been diagnosed with dementia and she really enjoyed getting out and spending time with other people. Margo and her sister both found that the program really helped them mentally and physically.

Alisa Grigorovich is Assistant Professor of Gerontology in the Department of Recreation and Leisure Studies at Brock University, Canada, and Affiliate Scientist at KITE Research Institute, Toronto Rehabilitation Institute – University Health Network. Her interdisciplinary research program centres on exploring social, ethical, and policy issues related to health, ageing, technology, and care. A central focus of her research concerns the development and evaluation of arts-based and digital initiatives to challenge stigma associated with dementia and to improve health and social inclusion across community-based and long-term care environments. Alisa was a postdoctoral collaborator on the Canadian Institutes of Health Research (CIHR)/Alzheimer Society of Canada *Improving Social Inclusion for People with Dementia and Carers through Sharing Dance* project.

Rachel Herron is Associate Professor in the Department of Geography and Environment at Brandon University, Canada, and the Canada Research Chair in Rural and Remote Mental Health. Her current research examines the vulnerability and complexity of care relationships, social inclusion, and meaningful engagement for people living with dementia and the diversity of lived experiences of rural mental health. She is the founding Director of the Brandon University Centre for Critical Studies of Rural Mental Health and was the co-principal investigator (with Mark Skinner) of the Canadian Institutes of Health Research (CIHR)/Alzheimer Society of Canada *Improving Social Inclusion for People with Dementia and Carers through Sharing Dance* project.

Jenn Kairies has over 20 years of experience making a difference in long-term care. With a background in Recreation Therapy and Leadership Development, she has successfully led and mentored teams of recreation professionals, behavioural support teams, and volunteers in providing older adults every opportunity to lead active and engaged lives while enjoying an inclusive and social atmosphere where they can truly feel fulfilled. Jenn's passion is empowering residents, promoting independence, and providing unique and creative ways to engage in past interests and hobbies. In addition to her role as a Recreation Supervisor and Educator in long-term care, recently Jenn has had the opportunity to share her passion and love for seniors and senior living in an education role, teaching courses in the Leadership in Senior Living Course offered at Conestoga College. Jenn was a community partner on the Canadian Institutes of Health Research (CIHR)/Alzheimer Society of Canada *Improving Social Inclusion for People with Dementia and Carers through Sharing Dance* project.

Jennifer Killing is Vice President, Quality, Research and Strategic Partnerships for peopleCare. She is a respected healthcare leader with over 20 years in senior leadership roles. As a member of peopleCare's senior leadership team, her emphasis is on quality initiatives and technology to promote exceptional resident care and outcomes. Recognized as an expert in her field, she is a frequent speaker and panelist at health system conferences, and regularly provides input at various strategic advisory tables. She is a board member of the Centres for Learning Research and Innovation (CLRI) in Long-Term Care as well as a member of the Ontario Long-Term Care Association Quality Committee, Institute for Safe Medication Practices (ISMP) Canada LTC Advisory Panel, and Canadian Advisory Committee member for pointclickcare. A certified Long-Term Care Administrator and RAI-MDS credentialed, Jennifer was a community partner on the Canadian Institutes of Health Research (CIHR)/Alzheimer Society of Canada *Improving Social Inclusion for People with Dementia and Carers through Sharing Dance* project.

Pia Kontos is Senior Scientist at KITE Research Institute, Toronto Rehabilitation Institute – University Health Network and Professor in the Dalla Lana School of Public Health at the University of Toronto, Canada. In her research, she focuses on stigma associated with dementia, the development of theories, policies, and practices that support relational caring, and draws on critical participatory and arts-based (visual arts, music, theatre, film) methodologies to reduce stigma and improve social inclusion and quality of care in institutional and community care settings. She was co- investigator on the Canadian Institutes of Health Research (CIHR)/ Alzheimer Society of Canada *Improving Social Inclusion for People with Dementia and Carers through Sharing Dance* project.

An Kosurko is Research Associate with the Trent Centre for Aging & Society at Trent University, Canada, where she recently graduated from the Master of Arts in Sustainability Studies Program. Her areas of research interest include ageing communications and technology, arts-based research, and community-based research. She is a Doctoral Candidate in the Ph.D. in Social Sciences Program and Researcher at the University of Helsinki, Finland. She was a Research Associate collaborator on the Canadian Institutes of Health Research (CIHR)/Alzheimer Society of Canada *Improving Social Inclusion for People with Dementia and Carers through Sharing Dance* project.

Verena Menec is Professor in the Department of Community Health Sciences at the University of Manitoba, Canada. She has been working in the area of age-friendly communities since 2006, when she conducted the focus group research for the WHO Age-Friendly Cities project in one of the participating cities in the province of Manitoba. She was also part of the research team for the Canadian Age-Friendly Rural and Remote Communities project. As part of a five-year program of research, she subsequently conducted many projects on age-friendliness in partnership with the Province of Manitoba's Age-Friendly Manitoba Initiative. She was a co-investigator in the Canadian Institutes of Health Research (CIHR)/

Alzheimer Society of Canada *Improving Social Inclusion for People with Dementia and Carers through Sharing Dance* project.

Sheila Novek is a Postdoctoral Fellow with the Safe Places for Aging and Care Project at Brandon University. She recently completed her Ph.D. in Community Health Sciences at the University of Manitoba, Canada. Her research examines the lived experiences of dementia, care relationships and contexts, and status and social inclusion of people living with dementia. She was a postdoctoral collaborator on the Canadian Institutes of Health Research (CIHR)/Alzheimer Society of Canada *Improving Social Inclusion for People with Dementia and Carers through Sharing Dance* project.

Mark Skinner is Dean of Humanities and Social Sciences at Trent University, Canada, where he is also Professor of Geography. A leading rural gerontology scholar, he is the past Canada Research Chair in Rural Aging, Health and Social Care and was the founding Director of the Trent Centre for Aging & Society. His previous Routledge books include *Ageing Resource Communities: New Frontiers of Rural Population Change, Community Development and Voluntarism* (2016), *Geographical Gerontology: Perspectives, Concepts, Approaches* (2018), and *Rural Gerontology: Towards Critical Perspectives on Rural Ageing* (2021). He was co-principal investigator (with Rachel Herron) of the Canadian Institutes of Health Research (CIHR)/Alzheimer Society of Canada *Improving Social Inclusion for People with Dementia and Carers through Sharing Dance* project.

Ruth Snider participated in the first remote pilot of the Sharing Dance Older Adults program in Ennismore, Ontario. Originally from the Newfoundland, Ruth Snider has a 35-year plan to live until 105 years old. She has diverse and international experience in Education and has been the Principal for adult programs, primary schools, and secondary schools. Ruth co-introduced native curriculums at the Adult Education Centre, Elliot Lake, and North Shore Board of Education and was on the boards of Work Force Development, Women's Crisis Centre and Northeastern Ontario Literacy Network. She also co-authored the best-selling book, "Teaching Styles, Learning Styles – Celebrating Differences". As the positive and lively spirit she is, Ruth encourages us to all: "Live the best life you can."

Melisa Stevanovic is Senior University Lecturer in the Faculty of Social Sciences at Tampere University, Finland. She has a Ph.D. in sociology from 2013, in which she examined power and authority during workplace meetings by using conversation-analytic methods. Thereafter she has conducted interdisciplinary studies on the embodied and "musical" aspects of joint decision-making interaction, making use of both naturally occurring interactions and naturalistic experimentation with motion capture, eye tracking, and physiological measurements. Her articles have been published in many peer-reviewed journals including Research on Language and Social Interaction, Text and Talk, Discourse Studies,

Social Semiotics, Social Psychology Quarterly, Psychology of Language and Communication, Scandinavian Journal of Educational Research, Journal of Pragmatics and Language in Society. She was an international collaborator on the Canadian Institutes of Health Research (CIHR)/ Alzheimer Society of Canada *Improving Social Inclusion for Older Canadians with Dementia and Carers through Sharing Dance* project.

Whitney Strachan is a Regional Coordinator at the Alzheimer Society of Manitoba in Brandon Manitoba, which hosted Canada's National Ballet School's Sharing Dance Older Adults program for older adults living with dementia and their carers. Whitney was a community partner on the Canadian Institutes of Health Research (CIHR)/Alzheimer Society of Canada *Improving Social Inclusion for People with Dementia and Carers through Sharing Dance* project.

Craig Wingrove is a musician and teacher who graduated from the Royal Conservatory of Music with his ARCT in piano performance and has his Honors BFA in music from York University. He has been a full-time musician at Canada's National Ballet School (NBS) for 30 years and a regular guest pianist for the National Ballet of Canada as well as at Arts Umbrella in Vancouver during the summer. He is one of the foremost pianists for the Cecchetti method and has been flown all over Canada to play for examinations, workshops, and teacher courses. He has produced 26 albums of dance class music, which are widely used all over the world, with more than four titles ranking top 20 consistently in iTunes. Craig also teaches all three levels of the Teacher Training Program at NBS and has written two books on music for the dance teacher. He is also the musician and composer in the pioneer team who developed Sharing Dance Older Adults at NBS, and regularly gives workshops to teachers and facilitators at Baycrest. He was a partner in the Canadian Institutes of Health Research (CIHR)/Alzheimer Society of Canada *Improving Social Inclusion for People with Dementia and Carers through Sharing Dance* project.

1 Introduction

Rachel Herron and Mark Skinner

Introduction

The arts – that ever-expanding array of creative practices, expressions, and performances – have always been fundamental to human flourishing; this is particularly true in relation to the life enrichment and well-being of older people in the 21st century. Indeed, there is a longstanding and growing interest in building knowledge about the arts as a means of experiencing, representing, and understanding ageing, both individually and collectively. As the best endeavours to understand human flourishing tell us, building arts-based ageing knowledge is, enduringly and often transformationally, a collaborative approach involving artists, scholars, and practitioners as well as older people themselves and their carers. This book is the story of one such collaborative endeavour.

The book focuses on the *art of dance* and the ways in which dance can enrich older adult lives. We focus on dance because it can offer people of all ages opportunities for creative expression, playfulness, laughter, learning and growth, social engagement, community building, inclusion, and movement (Chappell et al., 2021; Kontos et al., 2021). This list of benefits is far from exhaustive, but it clearly illustrates the holistic impacts of dance. In simple terms, we know dance feels good. It is good, not only for individuals but also for groups and communities. Despite the benefits of dance, it is often viewed as merely supplemental in ageing and health research as well as policy and practice (Herron et al., 2019; Kontos and Grigorovich, 2018).

Older adults are often denied the right to participation in dance as well as other arts-based and creative activities fundamental to human flourishing and life enrichment (Steele et al., 2020). There are many factors that prevent older adults from participating in dance. For example, ageist and ablest attitudes toward older adults can influence what researchers, policy-makers, and organizations think older adults can do as well as what outcomes or ends are desirable; this is particularly the case for older adults living with complex chronic conditions such as dementia (Shakespeare et al., 2019; Swaffer, 2015). In turn, these attitudes can influence what opportunities are prioritized and invested in as well as how older adults see themselves and what activities make them feel safe. It is also the case that

DOI: 10.4324/9781003260691-1

many older adults live in homes and communities where opportunities to engage in the arts are not accessible. There can be a lack of resources such as education and training, infrastructure and space, transportation options and mobility support, and staff and volunteers to carry out arts-based programming close to older adult homes, whether living in the community or residential care environments. What these barriers highlight is the social exclusion of older adults from arts-based and other activities. In addition, the multifaceted barriers older adults face point to the need for investment and collaboration to support arts-based programs across different settings.

In general, collaboration is critical to addressing the multifaceted challenges and opportunities of ageing, health, and well-being. In the case of the research that forms the basis of this book, collaboration was defined by working with different groups and individuals who have a vested interest in advancing opportunities for older adults, the arts, health and social care, and community development. All stages of the research involved coming together with different individuals and groups to make decisions about the design, promotion, delivery, data collection, and dissemination of the research (Skinner et al., 2018). Our transdisciplinary team included researchers with different disciplinary backgrounds (e.g., arts, health, and social sciences), and innovative perspectives from outside the academy, including the expertise of Canada's National Ballet School (NBS) and organizations involved in health and social care for older adults in community and long-term residential care (LTRC) settings. Most notably, the book itself is a creative collaborative endeavour featuring co-author contributions from the perspectives of older adults, family carers, health and social care providers, artists, and organizational leaders involved in the research.

As our approach attests to, focusing on dance, ageing, and collaborative arts-based research is timely and relevant for various groups. For older adults, family carers, and care providers, this book provides rich examples that illustrate the potential of arts-based programs and ageing research to amplify voices as well as enrich opportunities in later life. For the research community, this book contributes to a growing international literature on the arts, ageing, health, and care (Fancourt and Finn, 2019; Pedrazzani et al., 2017) as well as collaborative approaches to promote the health and well-being of older adults (Armstrong and Lowndes, 2018; Miller, 2021). The contributors to this book offer approaches, critiques, and alternatives to understanding arts-based programming as well as to advancing the concepts of social inclusion, relational dementia care, and embodied expression. The book also fills a need for applied perspectives that acknowledge the various factors communities and organizations must navigate in trying to make arts-based programming a reality. In particular, there has been increasing academic and practitioner interest in the use of technology to deliver arts-based programs in general as well as provide a broad range of supports for older adults (Kosurko et al., 2022). The contributors of this book explore the consequences of using technology to connect older adults. The book draws together each of these theoretical and applied strands of inquiry with the aim of advancing collaborative arts-based ageing research.

The goal of this introductory chapter is to present the distinctive aims of the book; outline the research project and specific objectives that provide a foundation for its contents; and introduce the collaborative approach that has guided this work. We begin by introducing the reader to the Sharing Dance Older Adults program and the opportunity to engage in collaborative arts-based ageing research. We review the literature to which this research contributes and offer a critique of the growing interest in arts-based programs within health, ageing, and dementia care research. The chapter also provides an introduction to the central concepts that thread the chapters together as well as the questions and innovative solutions each chapter offers. We conclude with some reflections on the value and limitations of this work.

An opportunity

The book is the culmination of a nationally funded four-year collaborative research project entitled *Improving Social Inclusion for People with Dementia and Carers through Sharing Dance*. However, the opportunity for this project began prior to a successful application for funding from the Canadian Institutes of Health Research (CIHR) and Alzheimer Society of Canada, and extends well beyond the completion of the research, as attested to in the postscript Coda at the end of the book. The opportunity emerged from shared interests in advancing both arts and ageing fields. It began, in part, because of NBS' passion for sharing the joy of dance with everyone.

NBS is an internationally renowned arts-based organization and a leader in dance for health and well-being. Although NBS has traditionally been recognized by public audiences worldwide as producing outstanding elite dancers employed by national ballet companies worldwide, it has a long history of excellence in teaching the performance art at all levels as well as making dance accessible. Out of these longstanding interests, a suite of community dance programs was developed, including a program designed in partnership with Baycrest, a renowned geriatric hospital, LTRC facility and research institute, for older adults with a broad range of abilities. After developing and delivering an initial in-studio version of Sharing Dance in Toronto, Canada's largest metropolis where NBS is located, the school continued to affirm and expand their philosophy of dance as well as their educational and artistic vision. They began to explore opportunities to share their knowledge of the benefits of dance more broadly through research, program expansion, and digital delivery in non-metropolitan regional settings. At this time, the co-editors of this book were introduced, which led to the relationship between NBS and, initially, Trent University, specifically the Trent Centre for Aging & Society (TCAS), and later Brandon University, which provided a foundation for further collaborative arts-based ageing research opportunities.

To expand their community dance programming, NBS initially secured funding from the Public Health Agency of Canada (PHAC). Then, the CIHR

Institute on Aging announced a funding program in partnership with the Alzheimer Society of Canada focused on social inclusion of individuals with dementia and carers, a successful application to which made it possible to explore the potential of sharing dance with older adults more comprehensively. The research team that developed around the CIHR/Alzheimer Society of Canada funded project to study the Sharing Dance program, all of whom are contributing authors to this book, were intrigued by its potential to promote ageing, well-being, health, and care at different levels. Although each team member brought different interests and perspectives to the table, we shared an enthusiasm and excitement about what Sharing Dance could offer individuals living with dementia, older adults in general, community organizations, institutions, and society more broadly. We knew that opportunities to participate in social life and engage in meaningful ways with others were often lacking for older adults living with dementia in LTRC and community settings, especially in non-metropolitan and rural communities with fewer programs and resources (Clarke and Bailey, 2016; Fortune and McKeown, 2016; Herron and Rosenberg, 2017). We were excited that the program could provide opportunities for people living with dementia and their carers to express themselves and be creative at an individual level. We wondered how the program could expand opportunities and resources in community and institutional settings. More broadly, we knew there was emerging evidence that arts-based programs could challenge stigmatizing views of dementia (Dupuis et al., 2016). With these possibilities in mind, we came together as a team to explore both the processes and outcomes associated with the program. Because we imagined that Sharing Dance might offer different experiences to different groups, including challenges and limitations, we engaged in partnerships with people living with dementia, carers, health and social care organizations, LTRC facilities, and Alzheimer Societies across multiple jurisdictions.

Our approach to collaborative research and, indeed, to writing this book together, was characterized by positive, constructive, and creative teamwork. We were interested in exploring what and how Sharing Dance worked as well as what did not work so well and how challenges could be addressed effectively for different groups in different community settings. Rather than view collaboration as a challenge in evaluating the Sharing Dance program, we sought to understand and inform best practices in the Sharing Dance program from the perspective of different groups. In many ways, the research herein is the product of the synergistic effects of working together and building on the contributions of others that are achieved through creative collaborative approaches.

The project

The overarching goal of our *Improving Social Inclusion for People with Dementia and Carers through Sharing Dance* project was to explore the potential of NBS' Sharing Dance Older Adults program, particularly in relation to improving social inclusion of older people living with dementia and carers.

We set three primary objectives to guide this exploration: first, to examine the experiences of older people (including people living with dementia, carers, and volunteers) participating in the program as they relate to the multiple dimensions of social inclusion; second, to assess the effectiveness of the remote delivery of the program to enhance social inclusion processes and outcomes and; third, to identify the challenges of expanding the program (e.g., programmatic and contextual factors that influence the success or failure of the program) in terms of improving social inclusion.

To address these objectives, we designed and conducted qualitative sequential pilot studies in two non-metropolitan regions of Canada (Peterborough, Ontario and Brandon, Manitoba), the research design and protocol for which was published by Skinner et al. (2018). The regional pilot studies involved collecting multiple forms of qualitative data (i.e., observations, diaries, focus groups, interviews, and reflective researcher notes) in different community and long-term care settings in each of the study regions between 2017 and 2019. With ethics approval from the universities, institutions, and organizations involved in the collaboration, multiple participant groups were engaged with their informed consent including older adults, people living with dementia, cares, volunteers, and staff and administrators. Each of these groups was included in the research to understand the experiences, effectiveness, and expansion of the program. Through a combination of transdisciplinary approaches, diverse methods, and voices, we were able to hear and see the impacts of Sharing Dance in depth, as exemplified in the poignant opening foreword to this book co-authored by an older adult caring for her partner living with dementia who were both participants in the Sharing Dance program.

Collaborating with different partners and engaging with different participant groups enabled us to take into consideration issues of capacity building and sustainability. We were aware at the outset that traditional health intervention studies often start and stop with research. Partners and participants often do not have the resources to continue programming. Early in the research, we brought partners and prospective participants together in both the Peterborough and Brandon regions to engage them in questions about how the Sharing Dance program and pilot research should work. To the best of our abilities, we wanted to learn and build the resources and strategies required to carry out and carry forward art-based programs like Sharing Dance. For example, we worked with local community organizations to apply for funding for media equipment (e.g., laptops and TVs) to offer Sharing Dance remotely as well as other programs in their community. Problem-solving was critical to building capacity and promoting more sustainable results beyond the research project. We sought to advance research practice as well as the theories that grounded our approach.

Foundations of the book

The research that forms the basis of this book is situated within the fields of arts-based health research, social gerontology, and dementia care. There is a growing body of evidence across these fields that arts programs such as dance can contribute to health promotion, disease prevention, and the treatment and management of health conditions (Fancourt and Finn, 2019). For example, there is a significant body of research on the use of dance to prevent falling and improve balance, mood, cognition, behaviour, and physical functioning among older adults (e.g., Abreu and Hartley, 2013; Beauchet et al., 2020; Brown et al., 2021; Coubard et al., 2011; Noice et al., 2014; Tymoszuk et al., 2020). Much of this research focuses on older adults with chronic conditions such as Parkinson's disease and dementia. Although this research has made important contributions to how older adults are treated, particularly in LTRC settings, research in this area tends to view older adults as passive recipients of programs carefully designed for them, rather than acknowledging their agency and creativity in arts-based activities (Zeilig et al., 2019). Viewing arts as a treatment also reduces the arts to an instrument for biomedical use rather than seeing the arts as a part of a rich life, to which older adults have a right. Moreover, the majority of research on arts-based programs for older adults is situated within an interventionist paradigm and has a tendency to focus on measuring outcomes rather than understanding broader processes that support and enrich older adult lives.

A small and growing number of studies have focused on the critical role of arts in addressing ageism and stigma (Black and Lipscomb, 2017; Dupuis et al., 2016; Gubner et al., 2020; Kontos et al., 2020). In these studies, the arts are not simply an instrument to be used on older adults; instead, older adults use the arts to challenge discourses that focus on decline and loss in later life. For example, Dupuis et al. (2016) found their community-based art project with people living with dementia, family members, and visual and performing artists created space to challenge dominant assumptions about dementia and develop more supportive and caring relationships. More recent research on dance by Kontos et al. (2021) illustrated how the playfulness and creativity that narrative dance offers can challenge the stigma associated with dementia. Other studies in the field of dementia care have observed enhanced communication and connection between formal care workers and people living with dementia as a result of participating in arts-based programs (Guzmán-García et al., 2013; Young et al., 2016) and they have linked arts-based programs to improvements in person-centred and relationship-centred care (Palo-Bengtsson and Ekman, 2000). Importantly, these studies suggest that the arts are not just good for older adults' health, they also have the potential to transform social attitudes and relationships.

Given the social nature and impact of arts-based programs, there has been a growing interest in the potential of arts to enhance social inclusion and broader determinants of health in later life. Social inclusion is an outcome and a process that involves supporting the rights and meaningful participation of older adults in the social world. The concept has become a guiding value in global action plans on ageing and dementia as well as national legislation and policies. Research has shown that community-engaged arts programming can foster social inclusion by enabling older adults to connect in new ways (Moody and Phinney, 2012). Other studies have examined the ability of dance to enhance social inclusion by increasing the number of social contacts people develop as well as addressing feelings of loneliness and isolation (Hansen et al., 2021). There is, however, a lack of in-depth exploration of how programs actually support social inclusion as a multidimensional process that involves promoting opportunities and addressing barriers to participation beyond the individual level. In this book, we draw on and contribute to recent multidimensional conceptual frameworks of social exclusion and inclusion that outline broader social, emotional, economic, cultural, and political processes that influence social inclusion (Pinkert et al., 2019; Walsh et al., 2019). In doing so, we are attentive to changes in approaches to providing health and social care for older adults, which shape practices and programs in different countries, communities, and institutions.

Good health and quality care of older adults takes place within the context of relationships to family, friends, professional care workers, and the state and its institutions (Miller and Kontos, 2016). There have been fundamental shifts in the way older adults are viewed and treated in health and social care over the past two decades. A comprehensive review of all the changes influencing the care of older adults is beyond the scope of this chapter; however, there are two broad theoretical shifts that highlight the limitations of biomedical and individualistic models of care provision, which fail to appreciate the diversity and interconnectedness of human well-being. First, there has been an increasing interest in person-centred approaches to care that value the identity and interests of the individual (Kitwood, 1997) and second, in response to the limitations of person-centred care, relationship-centred approaches have developed that value the growth and maintenance of caring relationships (Nolan et al., 2002), particularly in the context of dementia care. Arts-based approaches are well-suited to supporting both diverse older adults' well-being and advancing relational approaches to care. Dance programs, in particular, can offer older adults opportunities to interact meaningfully with their bodies, using non-verbal forms of communication (Kontos and Grigorovich, 2018) and, depending on the type of dance, it can also provide opportunities for physical touch and connection. These features of dance highlight some of the ways it can facilitate both individual expression and relationship building.

Building on extensive theoretical and empirical work using arts-based approaches with older adults, Kontos et al. (2017) advocate for a new model of citizenship that extends relational approaches to care and addresses social and structural disadvantages that older adults living with dementia face. A central tenet of this model is the recognition of the importance of embodied expression as a fundamental source of self-expression, interdependence, and reciprocal engagement. Rather than see older adults with dementia as lacking and unable to participate in programs because of their cognitive impairment, this approach shifts attention to power and agency of human bodies. Through their bodies, people living with dementia can express their desires, agency, and meaningful engagement with the world. To support the personhood and meaningful participation of older adults in relation to others, their embodied expression must be supported by institutional policies, programs, and practices.

Taken together, the concepts of embodied expression, relational care, and social inclusion centre the rights of older adults to meaningful arts-based programming. Each of these concepts builds on one another and provides a more comprehensive understanding of the role of the arts in ageing and health. These concepts guide the critical questions and innovative solutions within the various chapters of the book.

Advancing the field

Drawing together a transdisciplinary team of 21 researchers, artists, older adults, and leaders in ageing and care, the book aims to advance arts-based ageing research in terms of theories, approaches, and practices. The desires and rights of older adults drive this aim and are integrated into the ten chapters as well as the Foreword (written by Missy Drummond with Dennis Drummond) and Coda that comprise the book. Following the introduction by Rachel Herron and Mark Skinner (Chapter 1) are two foundational chapters that outline innovative approaches to program development and research that shaped the book. In Chapter 2, Rachel Bar and John Dalrymple highlight the distinctive development, expansion, and principles of Sharing Dance. They provide a rich example of how arts-based organizations can collaborate with the health and social care sectors to develop opportunities that foster creative and artistic experiences while also prioritizing accessibility. In doing so, they capture critical advancements in both the principles and practice of collaborative community dance initiatives with older adults. In Chapter 3, Mark Skinner and Rachel Bar describe the unique arts-based collaborative research approach that emerged from the *Improving Social Inclusion for People with Dementia and Carers through Sharing Dance* project. In doing so, they expand current approaches to arts, ageing, and health research, much of which has relied on care staff and family carers to evaluate the impact of the arts for health in rather traditional ways (Parsons et al., 2017). In championing 'creative collaboration', Skinner and Bar illustrate the critical role of methodological adaptations and advancements in collaborative, multisectoral research involving diverse stakeholders.

These introductory, contextual, and methodological foundations inform the suite of six chapters that form the analytical core of the book, beginning with two chapters that illuminate key concepts underlying our collaborative transdisciplinary research and illustrate how our project findings can advance debates and approaches. In Chapter 4, drawing on critical theory related to the impact of the arts in older adults' health, Rachel Bar and Pia Kontos share their perspectives as researchers and an artist about how Sharing Dance offers alternatives to the reductionism of the art-for-therapy culture and centres on human flourishing. Building on Chapter 3, they also highlight the importance of reflexive qualitative methods to advance and enrich understandings of the impact of dance on older adult lives. In Chapter 5, Rachel Herron, Sheila Novek, and Verena Menec explore how Sharing Dance can advance conceptualizations of social inclusion. Empirically, the chapter builds on multidimensional models of social inclusion, illustrating how processes and resources across various settings can come together to support social inclusion of older adults.

The next four chapters of the book focus on the empirical and applied contributions of the research, starting with the voices and lived experiences of key partners and participants. In Chapter 6, the perspectives of pianist Craig Wingrove (from Canada's National Ballet School in Toronto, Ontario), Sharing Dance participants Ruth Snider (from Ennismore, in Peterborough, Ontario) and Margaret Dunphy (from Minnedosa, Manitoba), and community facilitator Whitney Strachan (from the Alzheimer Society in Brandon, Manitoba) are brought together to explore first-hand the impact of Sharing Dance on different groups. With Rachel Bar and Rachel Herron, the stories in the chapter advance gerontological understandings of the value of integrating the arts in everyday places, not only for older adults, but also for the communities and institutions that surround them. In highlighting the mutual benefits of arts-based programs the chapter also raises important points about the role of joy, learning, and social connection in sustaining arts-based programs.

In Chapter 7, Pia Kontos and Alisa Grigorovich explore how Sharing Dance can challenge the stigma associated with dementia and advance research on the fundamental role of the body in self-expression, creativity, and sociability. Using rich examples, they also illustrate how the arts can cultivate a relational environment and relational approaches to care that counter biomedical understandings of ageing, which are relevant for practitioners. In Chapter 8, An Kosurko, Ilkka Arminen, and Melisa Stevanovic explore how various groups responded to technological challenges (or, 'glitches') in the remote delivery of Sharing Dance. Bringing an international perspective on Sharing Dance from the University of Helsinki, Finland, the chapter expands key debates about the limitations and potential of technology in program delivery for older adults. In Chapter 9, Verena Menec, Mark Skinner, and An Kosurko explore the intersection of arts-based programs and community development. The often complex and complicated dimensions and dynamics of communities are frequently neglected in research on arts, ageing, and health. Drawing on insights from Sharing Dance in rural communities, this chapter offers practical considerations in this area.

In Chapter 10, Rachel Bar and Rachel Herron conclude the book by reviewing the theoretical, methodological, and applied contributions of the *Improving Social Inclusion for People with Dementia and Carers through Sharing Dance* project, and analyses and reflections thereupon offered in the chapters of the book. With an eye to further knowledge mobilization in creative collaboration, the concluding chapter discusses the next steps for future arts-based program implementation and arts-based collaborative ageing research.

In the postscript to the book, Sheena Campbell, Cassy Borth, Jenn Kairies, and Jennifer Killing provide their perspectives as LTRC sector leaders, illustrating the continued importance and impact of arts-based programs for older adults during the COVID-19 global pandemic. Focusing on the implementation of a further adapted version of Sharing Dance that followed the *Improving Social Inclusion for People with Dementia and Carers through Sharing Dance* project, the Coda speaks to the pressing need to recognize the role of the arts in supporting human flourishing in times of crisis and to continue to transform care for older adults.

Overall, the book illustrates the potential of dance to enable bodies, enrich relationships and community opportunities, and challenge institutional cultures. In advancing theories of embodied expression, social inclusion, and relational care, the contributing authors raise questions about how researchers, professionals, and community leaders support the experiences and desires of older adults. Furthermore, the chapters in the book point to promising practices and next steps for meaningfully engaging diverse older adults in collaborative arts-based ageing research as well as innovative arts-based programs. Together, the contributing authors advocate that it is imperative that policies support accessible arts-based programs to enhance ageing, social inclusion, and care. In addition, they offer practical considerations for expanding such programs through the use of technology in diverse geographical settings.

Concluding comments

Dance, Ageing and Collaborative Arts-Based Research offers a broad understanding of the processes involved in developing, exploring, and understanding the impacts of arts-based programs on ageing, inclusion, and care. We emphasize the critical role of creativity, collaboration, and technology in meeting the fundamental needs and desires of older adults as they age in different contexts. At the same time, we offer an important and timely critique of the growing interest in arts-based programs within health and ageing research. In doing so, our aim is to redirect attention to processes that support life enrichment and well-being including the importance of addressing the stigma associated with age-related conditions, supporting meaningful relationships and connections, and enhancing accessibility to meaningful activities among older people.

Our research collaboration, and this book, is not without limitations. For instance, some of the voices of people living with dementia that we would like to have included in the book were no longer able to speak to their experiences because of their progressive condition. In addition, in terms of intersectionality (i.e., the multiple social categories that people occupy including gender, race, class, age, and ability), more work is needed to advance the inclusion and rights of diverse older adults in arts-based programs that are culturally relevant. Missing from the book are voices from older adults and carers who identify as racialized and/or Indigenous as well as those who identify as lesbian, gay, bisexual, transgender, intersex, or queer (LGBTIQ), and we join the call for future work in arts for ageing and health research to address these intersectional considerations. We also acknowledge the importance of global perspectives on arts and ageing research that, other than the inclusion of insights from North America and Northern Europe, are for the most part absent from the book, and we highlight in particular the potential to engage collaboratively with scholars, artists, and organizations, not to mention older people and carers themselves, in communities of the Global South.

The lessons emphasized in this book regarding collaborative, safe, inclusive, and innovative opportunities for ageing and health will be evermore important moving forward from the COVID-19 global pandemic. The pandemic and associated public health measures have brought tremendous change to the ways people typically engage with their social relationships and go about their day-to-day activities (Colibaba et al., 2021; Herron et al., 2021). As a result, some older adults have experienced increased isolation and exclusion from participating in the world outside their homes or even outside a single room within a congregate living environment (Fraser et al., 2020). In many countries, high rates of infection, hospitalization, and mortality among older adults in LTRC have fueled longstanding calls to transform care. Collaborative art-based solutions can contribute to broader system-wide innovation and transformation that enriches the lives of diverse older adults internationally.

References

Abreu, M., and Hartley, G. (2013). The effects of salsa dance on balance, gait, and fall risk in a sedentary patient with Alzheimer's dementia, multiple comorbidities, and recurrent falls. *Journal of Geriatric Physical Therapy*, *36*(2), 100–108. https://doi.org/10.1519/jpt.0b013e318267aa54

Armstrong, P., and Lowndes, R. (2018). *Creative Teamwork: Developing Rapid, Siteswitching Ethnography*. Canada: Oxford University Press.

Beauchet, O., Bastien, T., Mittelman, M., Hayashi, Y., and Ho, A. H. Y. (2020). Participatory art-based activity, community-dwelling older adults and changes in health condition: Results from a pre–post intervention, single-arm, prospective and longitudinal study. *Maturitas*, *134*, 8–14. https://doi.org/10.1016/j.maturitas.2020.01.006

Black, K., and Lipscomb, V. B. (2017). The promise of documentary theatre to counter ageism in age-friendly communities. *Journal of Aging Studies*, *42*, 32–37. https://doi.org/10.1016/j.jaging.2017.06.001

Brown, C. J., Chirino, A. F. C., Cortez, C. M., Gearhart, C., and Urizar, G. G. (2021). Conceptual art for the aging brain: Piloting an art-based cognitive health intervention. *Activities, Adaptation & Aging*, *45*(1), 39–69. https://doi.org/10.1080/01924788.2020.1719584

Chappell, K., Redding, E., Crickmay, U., Stancliffe, R., Jobbins, V., and Smith, S. (2021). The aesthetic, artistic and creative contributions of dance for health and wellbeing across the lifecourse: A systematic review. *International Journal of Qualitative Studies on Health and Well-being*, *16*(1), 1950891. https://doi.org/10.1080/17482631.2021.1950891

Clarke, C., and Bailey, C. (2016). Narrative citizenship resilience and inclusion with dementia: On the inside or on the outside of physical and social places. *Dementia, 15*(3), 434–452. https://doi.org/10.1177/1471301216639736

Colibaba, A., Skinner, M., and Russell, E. (2021). Rural aging during COVID-19: A case study of older voluntarism. *Canadian Journal on Aging / La Revue Canadienne Du Vieillissement, 40*(4), 581–590. https://doi.org/10.1017/S0714980821000386

Coubard, O. A., Duretz, S., Lefebvre, V., Lapalus, P., and Ferrufino, L. (2011). Practice of contemporary dance improves cognitive flexibility in aging. *Frontiers in Aging Neuroscience, 3*, 13. https://doi.org/10.3389/fnagi.2011.00013

Dupuis, S. L., Kontos, P., Mitchell, G., Jonas-Simpson, C., and Gray, J. (2016). Re-claiming citizenship through the arts. *Dementia, 15*(3), 358–380. https://doi.org/10.1177/1471301216637206

Fancourt, D., and Finn, S. (2019). *What is the evidence on the role of the arts in improving health and well-being? A scoping review*. World Health Organization Europe, Health Evidence Synthesis Report, 67. https://apps.who.int/iris/bitstream/handle/10665/329834/9789289054553-eng.pdf

Fortune, D., and McKeown, J. (2016). Sharing the journey: Exploring a social leisure program for persons with dementia and their spouses. *Leisure Sciences, 38*(4), 373–387. https://doi.org/10.1080/01490400.2016.1157776

Fraser, S., Lagacé, M., Bongué, B., Ndeye, N., Guyot, J., Bechard, L., Garcia, L., Taler, V., and CCNA Social Inclusion and Stigma Working Group et al. (2020). Ageism and COVID-19: What does our society's response say about us? *Age and Ageing*. https://doi.org/10.1093/ageing/afaa097

Gubner, J., Smith, A. K., and Allison, T. A. (2020). Transforming undergraduate student perceptions of dementia through music and filmmaking. *Journal of the American Geriatrics Society, 68*(5), 1083–1089. https://doi.org/10.1111/jgs.16418

Guzmán-García, A. H. J. C., Hughes, J. C., James, I. A., and Rochester, L. (2013). Dancing as a psychosocial intervention in care homes: A systematic review of the literature. *International Journal of Geriatric Psychiatry, 28*(9), 914–924. https://doi.org/10.1002/gps.3913

Hansen, P., Main, C., and Hartling, L. (2021). Dance intervention affects social connections and body appreciation among older adults in the long term despite COVID-19 social isolation: A mixed methods pilot study. *Frontiers in Psychology, 12*, 546. https://doi.org/10.3389/fpsyg.2021.635938

Herron, R. V., and Rosenberg, M. W. (2017). "Not there yet:" Examining community support from the perspective of people with dementia and their partners in care. *Social Science & Medicine, 173*, 81–87. https://doi.org/10.1016/j.socscimed.2016.11.041

Herron, R., Skinner, M., Kontos, P., Menec, V., and Bar, R. (2019). Beyond therapy: Exploring the potential of sharing dance to improve social inclusion for people living with dementia. In R. Hunt and S. Atkinson (Eds.), *Geohumanities and Health* (pp. 57–70). Switzerland: Springer.

Herron, R. V., Newall, N. E. G., Lawrence, B. C., Ramsey, D., Waddell, C. M., and Dauphinais, J. (2021). Conversations in times of isolation: Exploring rural-dwelling older adults' experiences of isolation and loneliness during the COVID-19 pandemic in Manitoba, Canada. *International Journal of Environmental Research and Public Health, 18*(6), 3028. https://doi.org/10.3390/ijerph18063028

Kitwood, T. M. (1997). *Dementia Reconsidered: The Person Comes First*. Buckingham: Open University Press.

Kontos, P., and Grigorovich, A. (2018). Rethinking musicality in dementia as embodied and relational. *Journal of Aging Studies, 45*, 39–48. https://doi.org/10.1016/j.jaging.2018.01.006

Kontos, P., Miller, K. L., and Kontos, A. P. (2017). Relational citizenship: Supporting embodied selfhood and relationality in dementia care. *Sociology of Health & Illness*, *39*(2), 182–198. https://doi.org/10.1111/1467-9566.12453

Kontos, P., Grigorovich, A., Kosurko, A., Bar, R. J., Herron, R. V., Menec, V. H., and Skinner, M. W. (2021). Dancing with dementia: Exploring the embodied dimensions of creativity and social engagement. *The Gerontologist*, *61*(5), 714–723. https://doi.org/10.1093/geront/gnaa129

Kosurko, A., Herron, R., Grigorovich, A., Bar, R., Kontos, P., Menec, V., and Skinner, M. W. (2022). Dance wherever you are: The evolution of multi-modal delivery for social inclusion of rural older adults, *Innovation in Aging*, 2022, igab058, https://doi.org/10.1093/geroni/igab058

Miller, E. (2021). *Creative Arts-Based Research in Aged Care: Photovoice, Photography and Poetry in Action*. Abingdon, UK: Routledge.

Moody, E., and Phinney, A. (2012). A community-engaged art program for older people: Fostering social inclusion. *Canadian Journal on Aging*, *31*(1), 55–64. https://doi.org/10.1017/S0714980811000596

Noice, T., Noice, H., and Kramer, A. F. (2014). Participatory arts for older adults: A review of benefits and challenges. *The Gerontologist*, *54*(5), 741–753. https://doi.org/10.1093/geront/gnt138

Palo-Bengtsson, L., and Ekman, S. L. (2000). Dance events as a caregiver intervention for persons with dementia. *Nursing Inquiry*, *7*(3), 156–165. https://doi.org/10.1046/j.1440-1800.2000.00064.x

Parsons, J., Gladstone, B., Gray, J., and Kontos, P. (2017). Re-conceptualizing 'impact' in art-based health research. *Journal of Applied Arts and Health*, *8*, 155–173. https://doi.org/10.1386/jaah.8.2.155_1

Pedrazzani, C., Cavalli, S., Di Giulio, P., Eiholzer, H., Paolantonio, P., Fancourt, D., and Williamon, A. (2017). Art for ages: The benefits of musical experience on the health and well-being in nursing homes. *Innovation in Aging*, *1*(1), 879. https://doi.org/10.1093/geroni/igx004.3157

Pinkert, C., Köhler, K., Von Kutzleben, M., Hochgräber, I., Cavazzini, C., Völz, S., Palm, R., and Holle, B. (2019). Social inclusion of people with dementia – An integrative review of theoretical frameworks, methods and findings in empirical studies. *Ageing & Society*, 1–21. https://doi.org/10.1017/s0144686x19001338

Shakespeare, T., Zeilig, H., and Mittler, P. (2019). Rights in mind: Thinking differently about dementia and disability. *Dementia*, *18*(3), 1075–1088. https://doi.org/10.1177/1471301217701506

Skinner, M. W., Herron, R. V., Bar, R. J., Kontos, P., and Menec, V. (2018). Improving social inclusion for people with dementia and carers through sharing dance: A qualitative sequential continuum of care pilot study protocol. *BMJ Open*, *8*(11). https://doi.org/10.1136/bmjopen-2018-026912

Steele, L., Carr, R., Swaffer, K., Phillipson, L., and Fleming, R. (2020). Human rights and the confinement of people living with dementia in care homes. *Health and Human Rights*, *22*(1), 7. https://doi.org/10.3390/laws8030018

Swaffer, K. (2015). Dementia and Prescribed Disengagement™. *Dementia*, *14*(1), 3–6. https://doi.org/10.1177/1471301214548136

Tymoszuk, U., Perkins, R., Spiro, N., Williamon, A., and Fancourt, D. (2020). Longitudinal associations between short-term, repeated, and sustained arts engagement and well-being outcomes in older adults. *The Journals of Gerontology. Series B, Psychological Sciences and Social Sciences*, *75*(7), 1609–1619. https://doi.org/10.1093/geronb/gbz085

Walsh, K., O'Shea, E., and Scharf, T. (2019). Rural old-age social exclusion: A conceptual framework on mediators of exclusion across the lifecourse. *Ageing & Society*, *4*(11), 2311–2337. https://doi.org/10.1017/s0144686x19000606

Young, R., Camic, P. M., and Tischler, V. (2016). The impact of community-based arts and health interventions on cognitive in people with dementia: A systematic literature review. *Aging & Mental Health*, *20*(4), 337–351. https://doi.org/10.1080/13607863.2015.1011080

Zeilig, H., Tischler, V., van der Byl Williams, M., West, J., and Strohmaier, S. (2019). Co-creativity, well-being and agency: A case study analysis of a co-creative arts group for people with dementia. *Journal of Aging Studies*, *49*, 16–24. https://doi.org/10.1016/j.jaging.2019.03.002

2 Canada's National Ballet School and the evolution of *Sharing Dance*

Rachel Bar and John Dalrymple

Introduction

Humans have been dancing for at least the last 40,000 years (Garfinkel, 2010). There are multiple reasons why people dance and countless ways to do so. The practice is shared across cultures and throughout human history, suggesting dance is adaptive to the human condition. The diversity and commonality of dance practices simultaneously differentiate and unite communities. While the attraction of many dance forms may be their unique and specific expression of movement to music, over the past few decades, the *community dance* movement has focused dance experiences around the communities participating in them.

Since the 1980s, there have been discussions and debates about the purpose and definition of community dance (Amans, 2017), but broadly the movement puts the group or community dancing together at the centre of its approach (People Dancing, 2021). Sharing Dance Older Adults is part of a suite of community dance initiatives Canada's National Ballet School (NBS) developed over the past decade. The program was designed to align with the principles that guide all of NBS' community initiatives as well as the School's mission as a national arts organization to share the transformative power of dance with excellence, access, and inclusion. From the very specific and defined nature of ballet to a diverse range of community dance offerings, this chapter outlines the unique journey and evolution that fostered the development of NBS' community dance initiatives and specifically Sharing Dance Older Adults. NBS is not the only arts institution to provide opportunity for community engagement, but this chapter highlights key features of NBS' approach which holds the art form at the heart of its practice, utilizes creativity and innovation, and prioritizes sustainability. Throughout this book, this approach and Sharing Dance's key features are revisited and considered from different perspectives.

In the beginning… A foundation for making dance accessible

NBS was founded in 1959 by two visionary, strong-willed, and accomplished women, Celia Franca, and Betty Oliphant. Both immigrants from England,

DOI: 10.4324/9781003260691-2

Franca was enticed to Toronto by the opportunity to establish the National Ballet of Canada; Oliphant at that time was regarded as one of the finest ballet teachers in the country. Seven years after the National Ballet was established, it was clear that Canada needed a national school as well. Established by Franca as the school of the National Ballet, Oliphant was appointed as the first Principal and Artistic Director, a position she would hold for 30 years. There were many principles underlying the School's foundation, but key among them were that talent would be the sole criterion for entry into the Professional Ballet Program; every dancer would receive a high-quality academic education under the same roof; and highest excellence would be the enduring programming standard. Franca and Oliphant also recognized that offering an onsite residence for students while attending the School located in downtown Toronto, Canada's largest metropolis, was critical for a national school serving a country as geographically expansive as Canada.

A single-minded approach to achieving world-class excellence has served the School well in its 60-plus years. Over time, NBS has matured into the largest professional arts training organization in Canada and is among the country's great arts and culture institutions. NBS graduates are performing in over 80 companies across the globe and NBS alumni are known for their leadership, directing major schools and dance companies around the world.

In the 1960s, NBS gained independence from the National Ballet of Canada in a move that legally severed the two entities in order to protect the real estate assets of the School's campus during a time of financial hardship. This type of operating model is extremely rare in the dance world, as the majority of schools are owned and operated by, and secondary to, ballet companies. It was perhaps NBS' first major expression of its creative and innovative approach to sustainability as an arts institution, but it would not be its last. NBS' independence has been the critical factor in its 60-year evolution, affording a kind of freedom to pursue a broader mandate and vision beyond what is needed in service of a company's repertoire or the preferences of its artistic leadership. It has allowed NBS to refine its brand to reflect a broader commitment to access and engagement in dance. It has enabled NBS to embrace a role in society that sees the excellence in ballet education that has been its cornerstone now benefitting a wider range of publics than ever before, regardless of age, ability, or aspiration.

While the School's independence has been a defining characteristic, it alone would not guarantee that NBS would be where it is today, as a leading ballet school globally with some of the broadest and most expansive range of community dance programming. So why NBS? While the School's Founders did not have a vision that included accessible dance programs for everyone in the community, their commitment to excellence and to talent being the only entry criterion for its Olympic-calibre training programs did set the stage.

Particularly the priority around talent ensured that no wealthy family could "buy" their child a place in the program and enabled aspirants to train at NBS regardless of their family's financial circumstances. The majority of students in the School have always required financial assistance to attend, as talent comes from all demographics and backgrounds. As Betty Oliphant's three-decades-long term as Artistic Director was coming to an end, a new generation of leaders was emerging – retired professional ballet dancers, still young, who were studying in NBS' Teacher Training Program. One such teacher was Canadian Mavis Staines, a graduate of NBS whose promising performance career was cut short by an accident. Staines was selected by Oliphant to be her successor. Staines, along with other colleagues like Laurel Toto and Anuschka Roes – who would go on to lead key areas of the School for decades – had all benefitted from accessible, community programs for dance and ballet. It was these community programs that had sparked their passion for ballet, and it was the financial assistance they received throughout their training that made their careers possible. As they stepped into leadership roles as educators, they were convinced that NBS should find ways to make the life-changing benefits of dance accessible to all (Figure 2.1).

When Mavis Staines became Artistic Director in 1989, she had a lot on her plate with regards to her commitment to reform elite ballet training at NBS and to influence change globally. While the traditional approach to training got results, it often came at a high cost. The training was intentionally harsh, and feedback often brutal. While those who persisted through the program developed deep wells of fortitude, the process also produced negative psychological outcomes, including high instances of distorted body image and disordered eating (Bar et al., 2017). Staines saw it as imperative that the

Figure 2.1 Dancers in the original studios at Canada's National Ballet School.

approach to training be overhauled, with a special focus on introducing health and wellness supports to the program, as well as an overhauled food service and nutrition program. Staines and her colleagues were also expanding the curriculum content and the breadth of the School's international partnerships. With excellence as the hallmark, the aim was to make every aspect of the School increasingly progressive.

It took decades for lasting, systemic changes to take place; for all of that time recognition of the importance of bringing dance to the community was never lost. However, necessary resources for the work were, for a time, elusive. Dedicated ballet teachers kept the dream alive on the sides of their desks. When they could carve out time, they would introduce creative movement programming to local elementary schools and began to expand now very successful on-site recreational ballet programs for children and adults.

50 years on… A breakthrough moment

As NBS approached its 50th anniversary in 2009/10, a series of celebrations were planned, including an unprecedented international gathering of the world's top ballet schools hosted in Toronto. As management developed further anniversary plans, a "gift" for NBS' recreational program participants was envisaged. This was 2009, when flash mobs were all the rage. While tirelessly choreographed and rehearsed, these public performances at train stations, malls and other crowded spaces appeared spontaneous to the delight of the surprised on-lookers. Videos capturing these fleeting happenings were garnering millions of views on YouTube. The 50th anniversary seemed the perfect moment to organize an NBS flash mob. Over 300 people from NBS' recreational programs committed to learning a joyous piece of choreography set to Leslie Feist's hit song "I Feel It All." Arrangements were made to perform at the Eaton Centre, Toronto's iconic downtown mall on April 29, 2010, International Dance Day. It was great fun and a big hit. No one on the day yet realized that this initiative would provide the "a-ha" moment that would drive the School's strategy for the next decade to come (Figure 2.2).

Once the flash mob was done and dusted, some key realizations emerged: 1) this approachable, fun engagement with dance had motivated hundreds of people with no aspirations or misapprehensions of a professional dance career to move their bodies to music, receiving all of the joy and benefits that dance offers; 2) the digital instructional videos that NBS created to help people practise were a proof of concept that NBS teachers did not need to be in the room to facilitate people dancing – the technology existed to do this at scale; and 3) the video of the flash mob was viewed online by more than 200,000 people – while this is modest compared to Canadian international superstar Justin Bieber's multi-billion video views at the time, it was significantly more engagement than the School's previous video views, which often numbered in the dozens or low hundreds.

At the same time that management at NBS was beginning to reckon with the potential of those realizations, the School was facing cuts to government

Figure 2.2 Canada's National Ballet School flash mob at the CF Toronto Eaton Centre, April 29, 2010.

funding and fundraising results that were stagnating. As the cost to deliver programming increased steadily each year, revenue was not keeping up. Facing a growing annual income gap, Staines was intent on developing a strategy that would see the School expanding its societal relevance and impact as a means to broaden its base of support, both from public and philanthropic sources.

Concurrently, the co-author of this chapter, John Dalrymple was a member of the School's Development Office, the department responsible for fundraising. In his third year at NBS, Dalrymple was the Associate Director of Annual Giving; previously, he spent nearly a decade in the museum sector both as a senior fundraiser and Executive Director. Knowing firsthand the challenges the fundraising group were facing with a shrinking donor base and the niche appeal of elite ballet training as a case for support, Dalrymple, like Staines, was convinced that future vitality depended upon expanding NBS' impact and demonstrating the broader relevance of dance. Rather than freeze in this moment of turbulence, the School looked to its culture of creativity and innovation to move forward.

A dream without a plan is just a nap

In a pivotal meeting in the spring of 2013, Staines and Dalrymple laid the framework for a concept that would guide the School for years to come: *NBS Sharing Dance*. While "Sharing Dance" was the name of a pilot project that sought to build on the flash mob model from 2010, Staines and Dalrymple had a bigger, more encompassing concept in mind: that if NBS Sharing Dance could begin to tackle real, sweeping challenges in society – things that a very broad range of people care about and support financially – the School would attract significant funding from sources that previously never imagined contributing to a ballet school. And that doing so at scale in meaningful ways would make dance and ballet relevant to many more people, sparking engagement in the art form that would see new pathways to training open up for talented kids who previously never saw themselves as artists.

In an institution all about dance, everyone knows that opportunities to dance can support human flourishment at the individual and community level, but to help tell this story to those who have not lived the joy of the art form, NBS needed to be intentional about for whom and how dance could contribute. In establishing NBS' Sharing Dance imperative, the School would focus on priority issues within society and the role that dance could play.

Physical activity for children would be the most obvious place to start since the focus of all of NBS' work had until this time predominantly focused on children and youth. Similar to other high income countries in the Global North, 90 per cent of children in Canada do not get enough physical activity (Government of Canada, 2017). Poor physical health in youth is a key indicator of an unhealthy adulthood resulting in greater instances of chronic disease (Government of Canada, 2017). Dance has all the health benefits of any physical activity, and it can often motivate children to be physically active, especially those not motivated by sports.

Supporting integration of new Canadians was also identified early on as a relevant focus for NBS' work. Canada is the most diverse nation in the G8 (Statistics Canada, 2018). Increased immigration is a key part of the country's development strategy, and the Ministry of Immigration, Refugees and Citizenship Canada's 2019/20 Departmental Plan underscores the consideration given to people's experience when they arrive and their integration into Canada in a way that maximizes their contribution to the country and fosters increased intercultural understanding (IRCC Departmental Plan, 2019). Dance is universal and present across cultures. Moving to music builds community, promotes social inclusion, and can often overcome language barriers.

While these two priority areas of focus would be compelling on their own, they left out a significant and growing sector of the Canadian population. Fortunately, subsequent sections of this chapter will describe how NBS came to recognize the *opportunities dance holds for older adults*. As one of the countries at the leading edge of global population ageing, there are more people in Canada over the age of 65 than under 15 for the first time in its history (Statistics Canada, 2022). With longer life expectancies, accessible innovations are needed to help older adults age with confidence, maintain health and well-being, and human flourishment. Dance provides all the benefits of physical exercise, but at its core, it is a creative art form and there are no limits to creativity. It can therefore be adapted to be engaging for different abilities and stages of life, including older adulthood.

As community dance programming as a dream and a function at NBS had been under-resourced for decades, it was decided that a new business model was needed to foster this strategic direction and focus in these three key areas.

Taking an innovative approach, Staines elected to create an entrepreneurial "start-up" within NBS: the Department of Strategic Initiatives, initially a department of one, headed by Dalrymple. The aim of the department was to embrace an entrepreneurial ethos within an established arts institution, using an iterative approach to designing and delivering initiatives as well as a venture capital-style approach to raising money to grow the department and its activities, all with the aim of generating a net financial return on activities within an initial three-year timeframe. With investments from NBS operations, seed funding from philanthropists Joan and Jerry Lozinski, long-time NBS donors, and a small innovation grant, the department launched in May 2013. While recognized internationally as a leader in ballet training, NBS was once again using creativity and innovation to redefine its operations and support its long-term sustainability.

It was determined that NBS' community dance initiatives would aim to provide people of all ages and/or aspirations with opportunities to experience the joy of dance, through excellence, access, and inclusion. Initiatives would be developed and continue to evolve in collaboration *with* communities. This community/participant-centred approach is central to NBS programs as they seek to remove barriers for those who have historically and/or are currently underrepresented or excluded from the dance sector (e.g., racialized persons, persons living with disabilities, etc.). Research and evaluation would inform and guide NBS community dance initiatives to ensure that the work was facilitating optimal dance experiences that maximize human flourishing. These aims set the framework for the community dance principles that now inform all of NBS' community dance initiatives (see Table 2.1). These initiatives are continuously evolving, but broadly include multiple dance opportunities for children and youth, an annual and inclusive national dance festival known as Sharing Dance Day, and several dance opportunities for older people.

While not the principal purpose, attracting growing streams of revenue from new sources was a major driver for institutional buy-in, which helped prioritize community dance initiatives and build momentum around the strategy. Since 2013 NBS Sharing Dance initiatives have attracted nearly CAD$18 million in direct funding, including a multimillion-dollar named gift and a groundbreaking, five-year partnership with the Public Health Agency of Canada (PHAC). As the strategic initiatives matured over time, a dedicated staff complement grew, further increasing NBS' capacity to expand its impact in community dance. New, permanent departments for Community Dance, Research and Health, External Affairs, and Brand Engagement were formed in the wake of successful "special projects". The strategic imperative towards NBS Sharing Dance was fully baked into the organization.

Dancing into the ageing/wellness space

It was right around the time that Staines and Dalrymple were laying out the framework for NBS Sharing Dance in 2013 that the School was approached by NBS alumna and chapter co-author Rachel Bar, who upon retirement from her professional dance career was pursuing graduate studies in Psychology. Bar was part of a research team led by Joseph DeSouza, Associate Professor, Systems Neuroscience at York University, and invited NBS to participate in a Parkinson's disease–related study (Bearss and DeSouza, 2021; Bearss et al., 2017). It had been well established that regular dancing improved motor symptoms and quality of life for people living with Parkinson's. NBS was asked to join the project by providing a 12-week dance program so that participants could be studied, an invitation that was readily accepted.

In order to train NBS faculty to deliver the dance program, David Leventhal, Program Director and founding teacher of Dance for PD® at the Mark Morris Dance Group in Brooklyn, New York led an in-person training workshop at NBS. Dance for PD® offers internationally acclaimed dance teacher training and dance classes for people with Parkinson's disease in New York City, online, and through a network of partners and associates in more than 300 other communities in 25 countries. In Dance for PD® classes, participants are empowered to explore movement and music in ways that are enjoyable, stimulating, and creative.

The faculty response to the training, followed by the community response to the weekly classes convinced NBS that this foray into teaching a dance program that was primarily for older adults would not end after 12 weeks. As a result, NBS Sharing Dance Parkinson's became one of the School's first permanent community dance programs (expanding over time to include additional on-site classes, annual teacher training seminars, additional research projects and the creation of Dance for Parkinson's Network Canada). Buoyed by the success and impact of the NBS Sharing Dance Parkinson's program, the School was keen to expand the impact it could make on the lives of older adults, particularly those with physical and/or cognitive challenges who would especially benefit from dance but may not be able to access traditional dance offerings.

This led to an important connection between NBS and Baycrest. Located in Toronto as well, Baycrest is a global leader in geriatric residential living, healthcare, research, innovation, and education, with a special focus on brain health and ageing. In 2014, Baycrest and NBS committed to working together to develop the Sharing Dance program, which would meet the needs of older people across the full spectrum of cognitive and physical ability – and were particularly committed to developing meaningful programming to engage older adults living with dementia in long-term residential care (LTRC).

Baycrest and NBS' Sharing Dance Older Adult Program

When NBS and Baycrest began to imagine what high-quality dance programming for older people could look like, the partnership was guided by the same principles that had evolved to guide all of NBS' community dance initiatives (see Table 2.1). These principles focus on fostering creative and artistic experiences while prioritizing accessibility. This meant that rather than apply a therapeutic lens to our dance programs for older people, design began from the

Table 2.1 Guiding principles of Canada's National Ballet School Community Dance Programs

Principle 1: Foster Creative Expression	Dance encourages participants to explore their inner artist and develop social and emotional skills in an imaginative setting. Creativity and play are key drivers that contribute to participants' intrinsic motivation and satisfaction. Dance creates the conditions for energetic discovery, allowing participants the opportunity to explore the world on their own terms, contributing to each individual's enjoyment and to the collective soul-stirring and life-enriching joy that strengthens community bonds.
Principle 2: Support Health and Well-being	Dance supports participants' physical literacy, develops confidence and competence, and contributes to overall physical, emotional, and social health and well-being. Dance motivates participants to value and engage in physical activities across the lifespan to support lifelong goals around wellness.
Principle 3: Provide Inclusive Opportunities	Dance creates accessible opportunities for participants of all ages, lived experiences, and states of health. Programs meet community needs through constant evaluation and iteration, including collaborations that ensure the inclusion of unique perspectives and cultural diversity. Technology is integral to program delivery, helping to remove geographic barriers and supporting sustainable program scale.
Principle 4: Build Community	Dance training and resources build capacity within communities and among individuals to deliver high-quality dance programming that creates safe spaces and develops positive and meaningful relationships among participants, teachers, and leaders. Dance fosters social inclusion, promotes increased social engagement, and inspires participants to take risks and explore new opportunities.
Principle 5: Advocate for Community Practice in Dance	Empower the dance sector to better understand and invest in community dance. Establish national, multi-sectoral partnerships to deepen and diversify sustained support to make the benefits of dance more accessible to all Canadians.
Principle 6: Advance Critical Research	Through dance research, capture the inherent contributions that dance offers to individuals – mind, body, and soul – and their communities. Drive research initiatives that advance our understanding of promising practices in community dance. Employ evidence-informed approaches to program design, content, and evaluation.

position that dance can be enjoyed by all if opportunities are made accessible. NBS faculty knew from research and their own experiences of dance that it offered many physical, emotional, and social benefits for people and communities, so the approach focused on creating access to the art form rather than an interventionist model that would directly target these specific outcomes.

How does one go about creating artistic and creative dance experiences for older people? To begin, NBS and Baycrest focused on accessibility. Recognizing, as with any age group, the huge variability of older adults' physical and cognitive abilities, Baycrest and NBS decided to initially focus on program development for older people with the least amount of access to traditional dance opportunities (i.e., dance classes for the general population) due to physical and/or cognitive challenges. Initial pilots took place in the LTRC facility at Baycrest. This population included older adults with significant physical and cognitive challenges, with most participants living with dementia. Although several participants could stand, the artistic team determined the program would be taught entirely from a seated position from the outset. This was thought to be less confusing for those participating who were not sure if standing was safe for them. It also allowed participants to focus more attention on artistic and creative qualities as maintaining balance while dancing did not need to be factored in. From a seated position, the artistic team began to experiment with dance phrases that offered both artistic and physical engagement. Through a development period of two years, the classes began to take on some of the structure of a traditional dance class. For example, as with any dance class, the class begins with a "warm-up" dance. It then works through a series of dances focusing on different creative, expressive, and physical aspects. A traditional dance class would typically build intensity as the class progressed, but through the development stage, the artistic team learned that a structure which gently built intensity in the first half of the class, dipped in intensity in the middle of class, and then built intensity again, supported participants' ability to remain engaged throughout the class. While the artistic team initially designed a 30-minute dance class for participants in LTRC, it became clear that with this structure, participants were often able to engage for 45 minutes.

Importantly, the 13 dances that make up the program's protocol are not set choreographies (see description of protocol in Tafler et al., under review). Their descriptions in the protocol are meant as a guide for artistic and physical engagement through a class. This means that within different settings and even within different classes the choice of music, genres of dance, and creative narratives can vary depending on the participants. Choreography is not set in the classes but instead offered as inspiration for participants to consider. For example, if a dance teacher presented a dance about going apple picking, they would then proceed to suggest several different movements to reflect how that might be expressed physically. Participants would be invited to use as much or as little of these movements as they would like or offer their own movements. As this approach encourages variability of movement by the dancers in the class, participants are reminded there is no right or wrong way to dance or express themselves creatively. Further, participants are encouraged to interact and engage with other dancers that are participating with them. Connection through eye-contact and even touch when appropriate is

Figure 2.3 Community dancers in class at Canada's National Ballet School.

encouraged in the approach to reflect traditional dance experiences (e.g., social dances). To further support participants' opportunities to experience the many gratifying experiences associated with dance, dances are usually kept for several weeks before changing class material. This provides opportunity for participants to build familiarity and confidence performing a dance and resembles the confidence and often enhanced enjoyment dancers experience as they rehearse a dance in preparation for performance (Figure 2.3).

The role of the carer was also an important consideration from the outset of this program's development. Baycrest and NBS recognized that for some carers, the dance classes could provide much-needed respite, but for those who stayed with the person they were caring for during the dance classes, two clear opportunities were recognized. First, carers were able to enjoy the dance class as a full participant and experience the joy and benefits of dance for themselves. Because everyone is welcomed to participate in whatever way feels right to them, carers were not required to "correct" the person they care for or facilitate any specific movements. Second, if they were dancing along with the person they care for, there was opportunity for social interaction that could be qualitatively different from their usual carer and recipient roles. This opportunity for a carer is highlighted in Margo Dunphy's contribution to Chapter 6 in this book.

After two years of developing the program in Baycrest's LTRC facility, Baycrest and NBS began to train dance teachers who had not been part of the program's development. Initially, three dance teachers were trained, who went on to teach the program at other LTRC facilities in the Greater Toronto Area. While Baycrest and NBS continued to hold annual training courses for dance teachers, it became clear that both the rate at which teachers could be trained and the reality that dance teachers tended to live in larger cities in Canada, were limitations. To address this, NBS and Baycrest began to consider other ways to increase access to dance for older people across the country.

Expanding access

By 2016, NBS had had several opportunities to experiment with livestreaming dance content. The School had begun to livestream performances by its professional students and was partnering with the Mark Morris Dance Groups' Dance for PD® program in Brooklyn to livestream Parkinson's classes from NBS once every few weeks. As NBS began to consider if/how the program being developed with Baycrest might be able to reach more people through remote delivery, the artistic team was serendipitously introduced to this book's co-editor, Mark Skinner, which led to the relationship between NBS and Trent University that set the stage for the research project discussed in this book. This meant the first time Baycrest and NBS tested the remote delivery of Sharing Dance it was in collaboration with Community Care Peterborough in Ennismore, Ontario, a village in Peterborough County, not too far from Trent University. Unlike the livestreaming that had been done for NBS' Parkinson's program, there was no concurrent class running in the background. Instead, the dance teacher on the screen was teaching directly to the group of dancers in Ennismore with only one "back-up dance teacher" and an accompanying pianist next to them on the screen. Further details related to the evolution of the remote delivery models piloted in this project are discussed by An Kosurko, Ilkka Arminen, and Melisa Stevanovic in Chapter 8.

Before running the initial pilot classes (also known as 'dress rehearsals'), members of the collaborative team from NBS and Trent University went to Ennismore to meet community members and many of the people who would be participating in the pilot. These were older people in the community who were already participating in an adapted yoga class for older people. Meeting them and dancing with them, it was clear that the program that had been developed within an LTRC context would need to be adapted for more independent older people. At this moment in the program's development, the protocol was adapted into two versions. Sharing Dance *In Your Seat* would follow the protocol developed in LTRC. *On Your Feet*,[1] was the version of the protocol that was created for the Ennismore pilot. While the overall approach of this version is the same as *In Your Seat*, it was created as a longer class (50–60 minutes) and included both seated and standing options. The target user was still an older person who would not find traditional dance classes accessible due to physical and or cognitive issues, but one who was able to be challenged with more complex directions in the dance classes than *In Your Seat* participants.

Meeting the community also allowed the project team to refine the concept of utilizing a program *facilitator*. This was the person who would organize and support the class in person. While the dance teacher from NBS was leading the class on a large screen, it was important that someone be on-site to support the group of dancers in person. For the first pilot, the adapted yoga instructor agreed to facilitate the dance program. While she did not have a dance background, she was able to set up the class each week, support and navigate any safety issues, and encourage the dancers to engage with the class and each other. As the program evolved, a 30-minute online training video was developed for on-site facilitators of the program, to help situate them into the role (Tafler et al., under review).

At first, classes were livestreamed, consistent with other offerings NBS had in development. However, the research project quickly highlighted the limitations of this approach. With internet stability less reliable in rural areas in Canada, the dancers' experiences of the classes were disrupted through this approach. To address this, the program model adapted towards on-demand video-stream delivery of pre-recorded weekly content. This also meant that as the pilot expanded to more sites and across time zones, each community could participate on a day and at a time convenient for them. Filming of classes took place once a week at NBS, and throughout the development stage classes were offered as eight-week terms. Although the dance teachers could not see their eventual participants, as one would using a live virtual platform, it was important that the artistic team connected and responded to their experiences. To accomplish this, on-site facilitators were provided with an online feedback form to fill out each week. These forms would let the dance teacher know what different sites were enjoying most/least about the classes and even asked if there were any musical requests or events happening in their community that may inspire a dance in an upcoming class. While the artistic team could not always incorporate every request that would be received, the dance teacher and live musician leading the classes would try to incorporate requests whenever possible.

The research project took the piloting of the program from the Peterborough Region of Ontario to the Brandon-Westman Region of Manitoba under the supervision of this book's co-editor Rachel Herron. In Manitoba, the communities using the program included people living with dementia, either living at home or in LTRC facilities. This gave the project team the opportunity to pilot the video-stream version of *In Your Seat*. At the time of the first pilot of *In Your Seat*, there was no certainty that people living with dementia would find the video-streamed program accessible or enjoyable. To the project team's knowledge, dance delivered remotely to people living with dementia had never been done before. Yet, as this book will describe, it became clear that this delivery model was indeed accessible and enjoyable for many participants living with dementia.

Program development beyond the research project

As the video-streamed program was piloted in more communities through this research project, the offerings continued to evolve based on learnings to maximize accessibility for different community contexts. Expanding and offering the program to communities beyond this project and at a significant scale was a clear priority for Baycrest and NBS. Funding from PHAC and this research project had been integral to getting this program launched; in keeping with the entrepreneurial ethos that drove the growth of community dance programming at NBS to date, a multifaceted revenue generation strategy was developed to build on the grant investments that were winding down. When the research project transitioned its focus from Ontario to Manitoba sites, some Ontario communities struggled to sustain the program. At that point, NBS began to explore effective ways of helping communities sustain the program. In some instances, communities were supported by NBS to seek their own grant funding to underwrite the cost of space, facilitators, and the small fee for the program. In other communities, individual participants paid an accessible per-dancer fee for the classes. Still other communities/institutions found ways within their own budgets and resources to prioritize the program. NBS continued to pursue additional philanthropic and government support, with doors increasingly open given the program's track record and growth. Baycrest and NBS also developed a commercialization agreement that empowers NBS to distribute the program to LTRC and retirement organizations, as well as individuals through an app-delivered subscription model (Figure 2.4).

By the fall of 2019, Sharing Dance was being offered in 60 sites across Canada in community, LTRC, and retirement settings. While there had always been a plan to develop an at-home version for people who were not able to leave their homes, the COVID-19 pandemic made the development of this version of the program a priority. Sharing Dance At-Home was piloted in the summer of 2020. Like the group model versions piloted in Peterborough and Brandon, Sharing Dance At-Home offers the *In Your Seat* and *On Your Feet* versions, but in shorter 20-minute dance bursts to be enjoyed at home. More about how this subsequent version was piloted with a new partner (people-Care Communities) and, especially, the experience of Sharing Dance during the COVID-19 global pandemic is provided in this book's Coda (Campbell, Borth, Kairies, and Killing). Formal research on this version of the program is not yet available but significant uptake and preliminary evaluation suggest the at-home version has provided meaningful dance opportunities to older people throughout the pandemic. In April 2021, NBS and Baycrest officially launched the above-mentioned Sharing Dance Older Adult app (https://oa.sharingdance.ca/). The app is optimized to further simplify and maximize access to the Sharing Dance programs. So much has been developed, expanded, and refined since NBS began this work, and with learnings from experience, evaluation, and research, Sharing Dance will no doubt continue to evolve to meet the needs of the communities engaging with the program.

Figure 2.4 Sharing Dance Older Adults, *In Your Seat* and *On Your Feet.*

Concluding comments

Looking back at the history of NBS and the evolution that led to the development of NBS' community dance initiatives, it is clear that artist-led creativity and innovation have always been key ingredients to NBS' success and sustainability as an arts organization. All at the School are excited about the ways in which NBS has been able to share the excellence in dance teaching it has always represented with people who may not have otherwise been able to access the joy of dance. The programs and initiatives now being offered and led by NBS are more robust and diverse than ever before. To further inform the School's programs and approach and expand its impact in the fields of dance and health, NBS' Research Institute was recently established as only the second arts organization in Canada to receive a research institute designation by Canada's federal research funding agencies. Just as NBS' professional ballet roots inform the excellence that is offered through its community programs, those same community programs now inform the evolution of the School as a whole… so much so, that in 2019, *Sharing Dance* fittingly became NBS' official tagline.

Sharing Dance was born within the unique artistic context of a ballet school. This distinctly centred its approach to dance for older people around opportunities to experience the art form. The subsequent chapters of this book further explore and highlight how this approach was experienced by its dancers and their communities.

Note

1 The names of NBS' dance programs for older adults changed multiple times through the development period for several reasons (previous names included *Sharing Dance with Seniors* and *Sharing Dance with Active Seniors* and later Level 1 and Level 2 were used). Critical social scientists involved in researching the program raised concern with the use of "senior" when describing older people. Further, describing the program for "active" seniors was confusing for some as the term "active" created different expectations for different people. The shift to describing the program as Level 1 and Level 2 confused some to think the program progressed from one level to the next with practice, as opposed to representing two distinct versions of the program. At the time this chapter was written, Sharing Dance Older Adults offered as *In Your Seat* or *On Your Feet*, appeared to be well received by those engaged with the program.

References

Amans, D. (2017). *An Introduction to Community Dance Practice*. Houndsmills, UK: Macmillan International Higher Education.

Bar, R. J., Cassin, S. E., and Dionne, M. M. (2017). The long-term impact of an eating disorder prevention program for professional ballet school students: A 15-year follow-up study. *Eating Disorders*, *25*(5), 375–387. https://doi.org/10.1080/10640266.2017.1308731

Bearss, K. A., and DeSouza, J. F. (2021). Parkinson's disease motor symptom progression slowed with multisensory dance learning over 3-years: A preliminary longitudinal investigation. *Brain Sciences*, *11*(7), 895. https://doi.org/10.3390/brainsci11070895

Bearss, K. A., McDonald, K. C., Bar, R. J., and DeSouza, J. F. (2017). Improvements in balance and gait speed after a 12 week dance intervention for Parkinson's disease. *Advances in Integrative Medicine*, *4*(1), 10–13. https://doi.org/10.1016/j.aimed.2017.02.002

Garfinkel, Y. (2010). Dance in prehistoric Europe. *Documenta Praehistorica*, *37*, 205–214. https://doi.org/10.4312/dp.37.18

Government of Canada. (2017). How healthy are Canadians? https://www.canada.ca/en/public-health/services/publications/healthy-living/how-healthy-canadians.html

Government of Canada. (2019). Immigration, refugees and citizenship (IRCC) Canada departmental plan 2019-2020. https://www.canada.ca/en/immigration-refugees-citizenship/corporate/publications-manuals/departmental-plan-2019-2020/departmental-plan.html

People Dancing UK. (2021). https://www.communitydance.org.uk/DB/resources-3/what-is-community-dance

Statistics Canada. (2018). Immigration and ethnocultural diversity in Canada. https://www12.statcan.gc.ca/nhs-enm/2011/as-sa/99-010-x/99-010-x2011001-eng.cfm

Statistics Canada. (2022). In the midst of high job vacancies and historically low unemployment, Canada faces record retirements from an aging labour force: Number of seniors aged 65 and older grows six times faster than children 0-14. https://www150.statcan.gc.ca/n1/daily-quotidien/220427/dq220427a-eng.htm

Tafler, M., Bar, R. J., Kim, S., and Sodums, D. (under review). The development of Sharing Dance Older Adults: Baycrest and Canada's National Ballet School make dance accessible to older adults with dementia. Manuscript available from corresponding author: rbar@nbs-enb.ca

3 A creative collaboration in arts-based ageing research

Mark Skinner and Rachel Bar

Introduction

Arts-based research has been broadly defined as the integration of any art form, such as literary, performance, or visual arts, at any point in the research process (McNiff, 2008) and, indeed, arts-based approaches to research have become increasingly important for building knowledge about the experiences, representations, and understandings of ageing (Amans, 2012; Hatton, 2021; Houston, 2019; Kay and Hammond, 2018; Miller, 2021). It has also become increasingly evident that collaboration lies at the heart of arts-based approaches (Daykin, 2019; Foster, 2015). As this book argues, the most impactful arts-based ageing research often involves creative teamwork among artists, scholars, and practitioners as well as, and most importantly, older people themselves and their carers. This chapter contributes to this aim by introducing the collaborative arts-based ageing research project, *Improving Social Inclusion for People with Dementia and Carers through Sharing Dance*, that is the intellectual and empirical focus of the book.

The research project focused on the Sharing Dance Older Adults program, an arts-based ageing innovation from Canada's National Ballet School (NBS) for older adults including people living with dementia, and much of this book highlights what was learned from our collaborative research with older people and their communities engaged in dancing. In addition to studying the Sharing Dance innovation, the individuals and organizations involved in the program guided our collaborative approach to the research project. Collaborative approaches also informed the way in which this project developed from a concept to the capstone publication of this book. There are many different terms and approaches to collaborative research, including community-based research (Cacari-Stone et al., 2014; Minkler, 2005), participatory research (Cornwall and Jewkes, 1995), team-based research (Armstrong and Lowndes, 2018), and integrated knowledge translation (Nguyen et al., 2020) to name just a few. Each of these approaches involves partners outside the university participating in decisions about how, by whom, for whom, and to what ends research is conducted (Cornwall and Jewkes, 1995; Minkler, 2005). Although each of these approaches differs in terms of their orientation, origins, and degree of power-sharing with partners,

DOI: 10.4324/9781003260691-3

all of these approaches share a focus on developing meaningful and trusting partnerships over time, collaborative decision making, and learning, and applying relevant knowledge to improve particular processes and outcomes such as enhancing programs, practices, policies, or societal understandings (Nguyen et al., 2020). Funders, governments, and researchers alike increasingly acknowledge the value of diverse collaborative approaches to advancing knowledge and its positive impacts on the lives of diverse groups.

In the case of the collaboration outlined in this book, there was a broad range of formal and informal partnerships (e.g., older adults, people living with dementia, artists and arts-based organizations, family carers, health and social care organizations, long-term residential care (LTRC) facilities, and Alzheimer Societies) involved in the research and their participation in various phases of the project varied based on their interests, time commitments, and abilities. Partners were involved at the outset, including in the process of applying for funding, planning the rollout of Sharing Dance across pilot study regions, and planning the appropriateness and usefulness of program promotion materials as well as research instruments. All partners and collaborators also received knowledge user reports with preliminary research findings throughout the project, and partners from each of these groups were invited to participate in developing this book. Meaningful partnership engagement was not without challenges and limitations throughout this project, but it is a guiding principle of the collaborative arts-based ageing research in this book.

In this chapter, we first describe our collaborative research on Sharing Dance with particular attention given to research design and methods. This in-depth overview of the processes, pilot study sites, and participants involved in the research establishes the methodological foundation for the remaining chapters of the book that focus, in turn, on the most important perspectives, debates, and findings that emerged from our project. In an effort to also illustrate the importance and impact of collaborative approaches in research, we next present an initial set of learnings that were captured through our research team meetings, which occurred regularly and purposefully throughout the duration of the four-year project to discuss, reflect on and refine the project's development in a process similar to what Armstrong and Lowndes (2018) refer to as 'creative teamwork'. We found that collaborative arts-based ageing research has its own unique opportunities and challenges, and we offer these insights from our experience in *creative collaboration* to inform and inspire promising practices for future research in this area.

A collaborative arts-based ageing research project

As introduced in the previous chapters (Herron and Skinner, Chapter 1; Bar and Dalrymple, Chapter 2), the opportunity to collaborate with NBS to evaluate and inform the impact of their Sharing Dance program emerged in the mid-2010s at a time when NBS was considering expanding the reach of the program to enable more older people to access dance across Canada. Working with Baycrest, a global leader in ageing research and residential care for older adults, NBS had initially piloted a live in-person version of Sharing Dance and was exploring an online version of the dance program that would reach older adults from diverse backgrounds, living environments, and parts of Canada. To help expand the Sharing Dance program beyond its Toronto studios, NBS successfully secured funding support from the Public Health Agency of Canada (PHAC), but they also sought to engage leading researchers to begin evaluating the design, scope, and impact of the program expansion. NBS had seen first-hand the health, social, and emotional benefits of dance in their live classes and, when the Canadian Institutes of Health Research (CIHR) announced a grant opportunity in partnership with the Alzheimer Society of Canada focused on social inclusion of individuals with dementia and carers, they were keen to partner with book co-editor Mark Skinner, who at the time was Director of the Trent Centre for Aging & Society (Trent University) and Canada Research Chair in Rural Aging, Health and Social Care, and his network of leaders in community-based social gerontology to explore further the potential of Sharing Dance.

In partnership with NBS (contributing author John Dalrymple and co-editor Rachal Bar) and led by the book co-editors Mark Skinner and Rachel Herron, and contributing authors Pia Kontos and Verena Menec, a collaborative research team subsequently mobilized around what became the CIHR/Alzheimer Society of Canada funded *Improving Social Inclusion for People with Dementia and Carers through Sharing Dance* project (see Figure 3.1). The collaboration that guided the project would involve four university faculty research investigators, two postdoctoral fellows, one Ph.D. student and seven graduate research assistants working with NBS staff, six professional dance instructors, 16 community administrators, 16 program facilitators and 20 lived experience experts (in this case, people living with dementia and family carers on Alzheimer Society Client Advisory Committees). It is the very transdisciplinary nature of this collaborative group and their purposeful and, at times, impromptu interactions in decision making, knowledge exchange, and co-production where 'creative collaboration' emerged as a coherent approach for the arts-based project.

The overarching goal of the research collaboration was to explore the potential of the Sharing Dance program, particularly in relation to improving social inclusion of older people living with dementia and carers. Our objectives were, first, to examine the experiences of older people (including people living with dementia, carers, volunteers) participating in the program as they relate to multiple dimensions of social inclusion; second, to assess the effectiveness of the remote delivery of the program to enhance social inclusion processes and outcomes and; third, to identify the challenges of expanding the program (e.g., programmatic and contextual factors that influence the success or failure of the program) in terms of improving social inclusion. To address these objectives, we conceptualized, designed, and implemented qualitative sequential pilot studies of the Sharing Dance program in non-metropolitan regions of two Canadian provinces (see Skinner et al., 2018). The pilot studies focused on the remotely instructed delivery of Sharing Dance in

Figure 3.1 The research team.

Clockwise from top left: Rachel Bar, Pia Kontos, Rachel Herron, An Kosurko, Alisa Grigorovich, Verena Menec, Mark Skinner

various settings (from the community to the institution) and involved five phases of qualitative data collection through observations, diaries, focus groups, and interviews with program participants, people living with dementia, carers, facilitators, instructors, and volunteers as well as critical reflections among research investigators and knowledge users.

The pilot studies

With ethics approval from the universities, institutions, and organizations involved in our collaboration, sequential pilot studies of the Sharing Dance program were carried out in the non-metropolitan regions of Peterborough, Ontario and Brandon, Manitoba (see Figure 3.2). These regions were ideal study sites because their ageing populations, various urban, small town and rural settlements, and continuum of care settings for older adults (Statistics

Figure 3.2 The Canadian pilot study regions.

Canada, 2022) are typical of what is considered non-metropolitan Canada where NBS was seeking to expand the reach of its Toronto-based programming. These two regions were also selected due to their proximity to researchers based at Trent University (located in Peterborough) and Brandon University (located in Brandon) whose network of research partners, such as Alzheimer Society, Community Care, and LTRC organizations, were keen to participate in the Sharing Dance program.

The pilot studies were designed to run sequentially, beginning in 2017–2018 with the Peterborough Pilot Study in partnership with Community Care Peterborough, which focused on the *On Your Feet* version of Sharing Dance, followed by the Brandon Pilot Study in 2018–2019 in partnership with the Alzheimer Society of Manitoba, which focused on the *In Your Seat* version of Sharing Dance designed to be accessible to people living with dementia. As described in Chapter 2 (Bar and Dalrymple), these versions of the program facilitate dance for participants with different physical and cognitive abilities, and a strength of our pilot study design was that it allowed for the study of both versions iteratively, enabling challenges and opportunities for program enhancement to emerge and inform subsequent phases. In each of the two regions, the pilot studies were conducted in three phases starting with a 'dress rehearsal' phase at a single community site to establish and ensure technical, instructional and administrative capacity for the Sharing Dance program, followed by two subsequent phases of expansion to multiple community and institutional (LTRC) settings where the qualitative research was undertaken in full.

The *Peterborough Pilot Study* took place at multiple sites (seven in total) in the branch offices of Community Care Peterborough, as well as in a retirement residence, located throughout the region (see Figure 3.3). In the dress rehearsal phase (P1), the research team based out of Trent University, NBS, and Community Care worked collaboratively to test the first remotely delivered group session of the Sharing Dance Older Adults *On Your Feet* program. The eight-week session dress rehearsal took place in Spring 2017 at Community Care Peterborough's branch office in the village of Ennismore. The dress rehearsal was followed by a research collaboration planning and knowledge exchange meeting at Trent University in Summer 2017 (see report by Kosurko et al., 2017). In the community expansion phase (P2), additional Community Care Peterborough branch offices in Apsley, Ennismore, Havelock, Lakefield, and Millbrook participated in the Sharing Dance program, through weekly sessions in Fall 2017 and Winter 2018 (see report by Warrener et al., 2018a). The institutional expansion phase (P3) involved running the Sharing Dance program in the Community Care Peterborough branch offices and the William Place Retirement Residence located in Lindsay in Spring 2018 (see report by Kosurko et al., 2018).

The *Brandon Pilot Study* involved research at multiple study sites (six total) with several organizations and institutions located throughout the region known as Westman (see Figure 3.4). In the dress rehearsal phase (B1), the research team based out of Brandon University, NBS, and the Alzheimer Society of Manitoba branch in Brandon worked collaboratively to test the first remotely delivered group session of the Sharing Dance Older Adults *In Your Seat* program with people living with dementia over eight weeks in Winter 2018. The dress rehearsal was supported by a research collaboration planning and knowledge exchange meeting at Brandon University in Winter 2018 (see report by Warrener et al., 2018b). The community expansion phase (B2), which overlapped with the institutional expansion phase (P3), was hosted at three sites including the Prairie Oasis Senior Centre in Brandon, the Minnedosa 50+ Activity Centre, and Country Meadows Personal Care Home in Neepawa in Fall 2018 (see report by Kosurko et al., 2019). The institutional expansion phase (B3), concluded in Spring 2019 at five institutional settings and one community setting at the Country Meadows Personal Care Home in Neepawa, along with Personal Care Homes in Killarney, Minnedosa, Hamiota, and Carberry (see report by Kosurko et al., 2020a).

Qualitative research methods

To achieve our goal and objectives to better understand the experiences, effectiveness, and expansion of Sharing Dance, our sequential pilot studies

Figure 3.3 Peterborough pilot study sites.

involved five phases of qualitative data collection in the form of observations, diaries, focus groups, and interviews with program participants, including people living with dementia, carers, facilitators, dance teachers, and volunteers as well as critical reflections among research investigators and

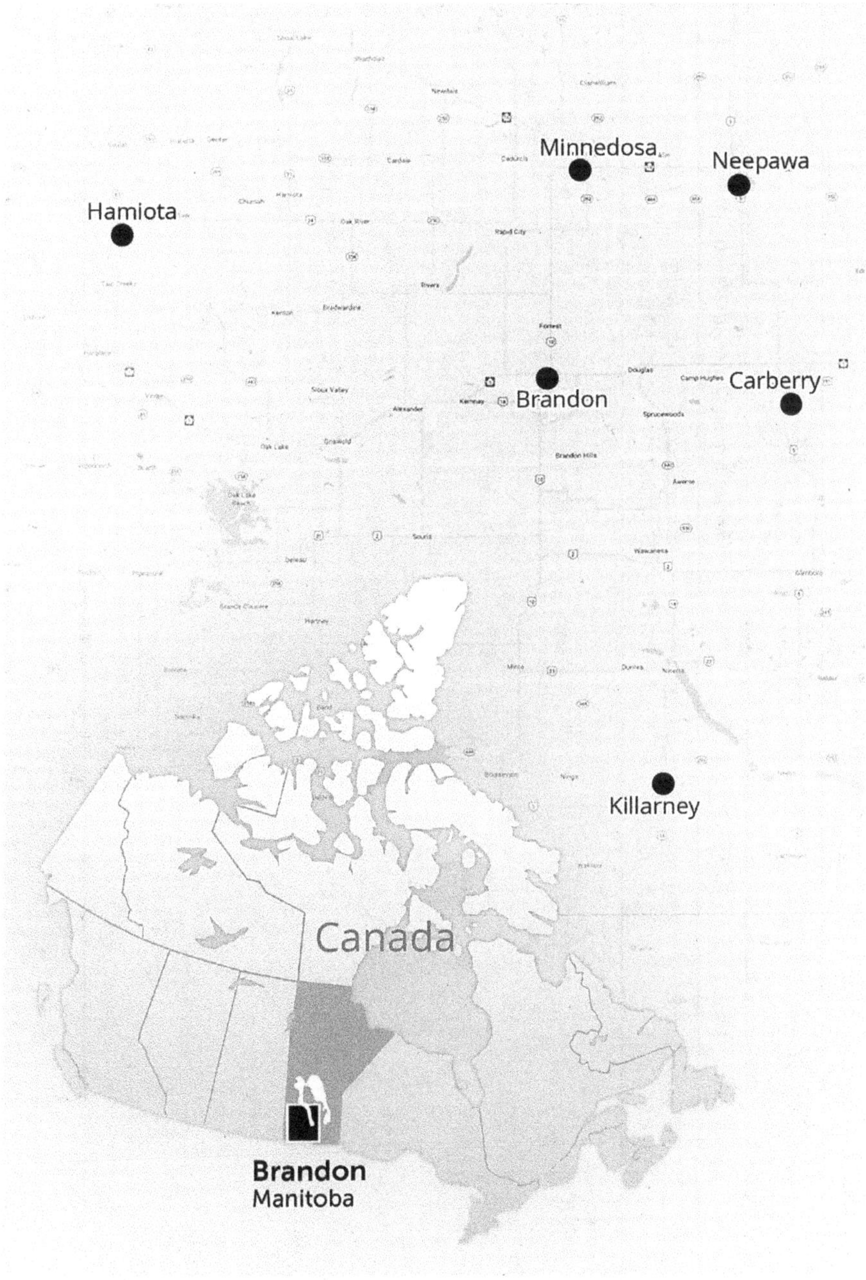

Figure 3.4 Brandon pilot study sites.

knowledge users. As summarized in Table 3.1, our qualitative research was designed to elicit and triangulate multiple forms of information (i.e., observations, experiences, perspectives, reflections) from different groups (i.e., older

Table 3.1 Qualitative sequential pilot study design

Research Design		*Peterborough Pilot Study (2017–2018)*			*Brandon Pilot Study (2018–2019)*		
Program Level		*Sharing Dance Older Adults* – On Your Feet			*Sharing Dance Older Adults* – In Your Seat		
Research Phase		*Dress rehearsal P1 (1 site)*	*Community P2 (5 sites)*	*Institution P3 (5 sites)*	*Dress rehearsal B1 (1 site)*	*Community B2 (3 sites)*	*Institutional B3 (6 sites)*
Methods	**Observations**						
	- Participants	23	54	40	11	20	122
	- Carers	—	—	—	5	16	11
	- Volunteers	—	—	—	—	9	7
	Diaries						
	- Participants	23	10	6	15	10	1
	- Carers	—	—	—	5	13	5
	Focus groups						
	- Participants	23	26	16	5	22	36
	- Carers	—	—	—	6	8	12
	- Facilitators	—	3	3	1	10	10
	- Volunteers	—	5	1	—	1	5
	Interviews						
	- Participants	23	13	17	3	12	19
	- Carers	—	—	—	3	4	4
	- Facilitators	—	2	3	2	8	10
	- Volunteers	—	—	—	—	1	3
	Reflections						
	- Investigators	6	6	6	6	6	6

Knowledge Exchange and Dissemination	**Phase 1 Report** (Kosurko et al., 2017)	**Phase 2 Report** (Warrener et al., 2018a)	**Phase 3 Report** (Kosurko et al., 2018)	**Phase 1 Report** (Warrener et al., 2018b)	**Phase 2 Report** (Kosurko et al., 2019)	**Phase 3 Report** (Kosurko et al., 2020a)
	Research Project Synthesis Report (Kosurko et al., 2021a) **Research Project Website, Presentations and Media** (www.sdsenionrsresearch.com) **Research Publications** (Herron et al., 2020; Kosurko et al., 2020b; Kontos et al., 2021; Kosurko et al., 2021b; Kosurko et al., 2022; Skinner et al., 2018)					

Adapted from Skinner et al. (2018, p. 4)

people with and without a dementia diagnosis, family carers, volunteer assistants, program facilitators, and administrators) in different settings (i.e., community care and institutional LTRC) (see also Kosurko et al., 2021a; Skinner et al., 2018). Across the two pilot studies, 289 participants were involved in the multiple phases of research, including 98 people living with dementia, 32 carers, and 16 volunteers. All of the participants provided informed consent individually or by proxy through a family carer, and the research team practised ongoing consent by monitoring participants' verbal and gestural cues throughout the program.

Observations involved detailed descriptions of the actions, expressions, and interactions of participants in Sharing Dance classes to inform our analysis of multiple dimensions of social inclusion, particularly those related to social relations, mobility, and safety. For example, observations focus on the degree and nature of social interaction as well as safety (e.g., comfort with dance moves, fellow participants, and the facilitator). Researchers and research assistants both observed and participated in the weekly Sharing Dance classes, writing detailed, concrete, and illustrative descriptions of participant interactions and expressions as they occurred over the course of the program. Written observations followed a loosely structured guide with prompts about the physical setting, participants' gestures, facial expressions and interactions with others, the sequence of activities and responses to the dance teachers on screen, and communication between the onsite facilitator and the participants. With participant consent, dance classes were also video-recorded in some sites to supplement written notes.

In addition to observing how people interacted in the program, participants in Sharing Dance were invited to complete a diary immediately after each dance session with the support of a research assistant or LTRC staff. *Participant diaries* involved the personal elicitation of the lived experiences of engagement with the program as the participants relate to processes and outcomes of social inclusion. People living with dementia and carers recorded their personal experiences throughout the program in a diary journal using written, oral, and/or video communication (paper booklets and iPads). In one LTRC residence, diary prompts were shared with the group of participants and volunteers, staff, and a researcher assisted them with documenting their responses. Although relatively few participants kept a diary and most of these participants were living in the community setting (i.e., not in LTRC), the few diaries that were completed complemented observations enabling participants to tell us more about what the research team saw in observations in their own words (see Herron et al., 2020).

Focus group discussions with program participants and carers as well as with program facilitators, instructors, and volunteers were conducted regarding the effectiveness of the program delivery and the challenges of implementing the program. This allowed for an examination of the breadth of experiential, programmatic, and contextual factors that influence the effectiveness of the program to improve social inclusion. Focus groups were conducted following the last dance session in each Pilot Study phase. Focus groups followed a semi-structured guide, which covered the effectiveness of the program and the challenges of participating and/or implementing the program.

Interviews were also conducted with older adults, carers, program facilitators, and volunteers to get richer individual accounts of individual experiences. Distinctive semi-structured guides and protocols were developed for older adults/family carers and facilitators. Older adults and carers were asked about their perceptions of the program, their dance experiences prior to the program, participation in other community activities, barriers to participation, and experiences of loneliness prior to the program. Then, they were asked to tell us about their experiences of the program; what they liked best; what, if anything, made it challenging to participate; what, if anything, had the program changed in their life; and what they would change about the program.

Interviews with program facilitators and volunteers were also undertaken, following the focus groups, to explore the effectiveness of the multi-modal program delivery and to identify the challenges of implementing the program. Semi-structured interview questions explored in greater depth issues and connections among the multiple dimensions of social inclusion that emerged from the focus group discussions. Interviews with staff and volunteers followed a semi-structured guide and focused on the effectiveness and challenges of implementing the program as well as the impact of the program on them and their organization. For example, they were asked about what aspects of hosting/facilitating the program they found most challenging (related to participants, instructions, the setting, resources, technology, and remote delivery) and they were asked if they could identify a particular moment or experience during the program that had a big impact on them. All interviews and focus groups lasted between 30 and 60 minutes, and were digitally recorded and transcribed verbatim.

Critical reflections among the research investigators and knowledge users served to identify challenges, opportunities, and promising practices for expanding the program in various types of community and collaborative settings. In-depth reflections regarding the challenges and opportunities experienced and/or observed by the research team (comprising the investigators, knowledge user collaborators from NBS, and research personnel) were recorded as part of weekly one-hour meetings to build a reflective dataset as the program expanded through the pilot phases. Detailed reflections were recorded in a text file as part of weekly one-hour team meetings that occurred regularly and purposefully throughout the duration of the four-year research collaboration. More than 100 pages of reflective notes were recorded over the course of our collaborative research.

Ethical participant engagement

Underlying our research design is a commitment to the safe and inclusive engagement of participants throughout the research process, especially those living with dementia (Novek and Wilkinson, 2019). This included developing relationships with advocacy organizations, considering and adapting language throughout the research process to maximize understanding and reduce stigma, mitigating distress and ensuring participant comfort, and continuous learning and adaptation among the research team members. As mentioned above, multiple Alzheimer Society organizations, a client advisory committee, and a dementia support group comprised of people living with dementia reviewed the study design, research instruments, and promotion prior to recruitment and had opportunities to make changes to materials and processes through this engagement. During the course of data collection, consideration was also given to participant comfort and choice. Participants could choose to participate in data collection that most suited them and they had choices about where, for example, interviews should take place and with whom (e.g., with or without a carer). As the next major section of this chapter attests to, learning and adaptation was a critical component of both our approach to ethics and collaborative research.

Data analysis

Following an iterative approach, analyses of focus group and interview transcripts, diaries, and observations took place concurrent with data collection. This process involved research team members reading through textual data and developing thematic codes through detailed line-by-line analysis within NVivo 11 software (Braun and Clarke, 2006). In each new phase of the data, codes from subsequent phases of data analysis were reviewed and new emergent codes were added and discussed by the team to develop a rich contextualized understanding of the themes across the Peterborough and Brandon pilot studies. This sequential and iterative approach enabled the research team and collaborators to learn and adapt from inductive, thematic findings as well as reflective notes as the study progressed. For example, and as discussed further in the next section, thematic content analysis of text-based observations informed not only subsequent interview and focus group prompts, but also the development of the Sharing Dance program. In addition, more detailed knowledge user reports and presentations were also used to share knowledge throughout the process of data collection (see Kosurko et al., 2018, 2019, 2020a, 2021a; Kosurko et al., 2017, Warrener et al., 2018a, 2018b).

The findings of our analysis, and our reflections thereupon, form the basis for the next suite of chapters in the book that emphasize, in turn, the most important theoretical and empirical contributions of the *Improving Social Inclusion for People with Dementia and Carers through Sharing Dance* project. Expanding on our initial publications from the project by Skinner et al. (2018), Herron et al. (2020), Kosurko et al. (2020b), Kontos et al. (2021), Kosurko et al. (2021b), and Kosurko et al. (2022), in an effort to illustrate the scope for and illuminate the importance of collaborative arts-based ageing research approaches, each of the six chapters that follow draws on, variously, project processes and findings to centre human flourishing within researcher-artist interactions (Bar and Kontos, Chapter 4), understand how to support social inclusion of older adults (Herron, Novek, and Menec, Chapter 5), hear and amplify voices of those experiencing the impacts of Sharing Dance (Bar, Dunphy, Herron, Snider, Strachan, and Wingrove, Chapter 6), challenge the stigma associated with dementia (Kontos and Grigorovich, Chapter 7), expand debates about the role of technology in delivering arts-based programs (Kosurko, Arminen, and Stevanovic, Chapter 8), and understand the community dynamics of dance and ageing (Menec, Skinner, and Kosurko, Chapter 9). Prior to these important and insightful contributions, however, we turn in the section below to a consideration of the role of researcher reflexivity through our *creative collaboration* in building knowledge about arts-based ageing research.

Building knowledge through researcher reflections

Implementing arts-based initiatives in communities is not for the faint-hearted – nor is the attempt to research them. To inform future work in this field, we used a reflexive and dialogical approach to consider what opportunities, challenges, and promising practices arise from our collaborative research project on Sharing Dance. To explore these themes, we draw on the critical reflections that were obtained from weekly team meeting notes. Right from the initial pilot study, weekly meeting agendas and notes were compiled into a single running document (totalling more than 100 pages of reflective notes over the course of our collaborative research), which were manually analysed thematically following Braun and Clark's (2006) approach. Ongoing discussions and analysis followed each meeting, allowing for a reflexive and dialogical approach to answering the questions above. The findings from our reflections are presented here, as both an articulation of the opportunities and challenges that emerged in the project and an illustration of applied collaboration among the leadership of the research team.

Opportunities

Through thematic analysis, reflection, and discussion of team meeting notes, the research team identified three key opportunities arising from our collaborative arts-based research on Sharing Dance.

Opportunity 1. Transdisciplinary learning. Our collaborative, arts-based research project provided ample opportunities for academics, students, artists, and community stakeholders to learn and inform practices in new and different fields. From the very beginning, this project embraced a transdisciplinary approach, meaning that knowledge and reflections of people outside of academia were included (Mobjörk, 2010). Further, the role of the researchers in this project was not restricted to data collection. Often the researchers danced along with participants and helped to organize and provide feedback about the program that informed the dance program's development. The research associate observing dance classes for the research study also regularly shared reflections, insights, and suggestions with NBS based on what they had observed in the dance classes. Conversely, one of the dance artists teaching classes was also part of the research team and contributed to the research project by adding the perspective of the artist to the research approach. Involving the artist in all aspects of the research project and not just the art-specific component of a project is not always how arts-based research is done. For example, Bartlett (2015) hired an artist to create an art installation about living with dementia but found the art overtook the research goals because the goals were not integrated. Community stakeholders also informed both the research and dance program's development by requesting both the research team and NBS provide a demonstration/explanation session before the start date in their communities to encourage participation among other suggestions. A collaborative arts-based research project is a great opportunity to learn and appreciate different perspectives and ways of knowing associated with different opportunities. When open to this approach, there is opportunity for insights from different disciplines and experiences to infuse and inform a project for optimal learning and outcomes.

Transdisciplinary learning also specifically facilitated the opportunity for the research project to inform the innovation. Our collaborative arts-based research provided significant opportunity for the research to inform the real-time development of the dance innovation (i.e., Sharing Dance). Feedback from the research team about observed participant experiences in the dance classes led to changes in how Sharing Dance was offered to communities. NBS asked "if all observations to date could be shared to help … understand how the program is going" (noted on February 16, 2018). Through the sharing of insights across disciplines, the research team was able to provide observation notes that informed the adaptation of content and presentation to maximize the accessibility of the program. Researchers often separate themselves from what is being studied to remain objective and not interfere, but in this project, the research-innovation interaction supported transdisciplinary learning that ultimately supported the research, the innovation, and the community.

Opportunity 2. Phased Research. When planning a four-year research project, it is tempting to try to collect as much data as possible early on while the excitement and momentum of a newly successful grant application is still present. For collaborative arts-based research, however, this project helped to identify why a phased approach may be most effective at maximizing opportunity for learning and meaningful outcomes. With each data collection phase of the project, new insights were gleaned that informed the project's research methods and the dance innovation. By collecting data in distinct phases with several weeks between the end of one phase and the start of the next, the research team was able to take the time to assemble preliminary reports that guided subsequent phases (see Table 3.1 earlier in the chapter). For example, research team meeting notes recorded, "We discussed the need to plan data collection processes to be less complicated in [the next phase] (noted on March 2, 2018)". When the research team reflected on the experience of participants in the second phase of the Peterborough Pilot Study (P2), we were able to consider the ways that data collection could be simplified to support a better experience for participants in subsequent phases.

As another example, team meeting notes captured reflections about research tools. Recognizing "The caregiver diaries were not helpful" (noted on March 16, 2018) in a team meeting helped to refine the data collection tools for subsequent phases of the study. This saved project resources and allowed the team to focus time on collecting and analysing the other types of data collected. This collaborative arts-based research on Sharing Dance project had multiple logistical considerations that took time to coordinate in the multiple communities involved. The benefit of the coordination work taking time was that it also provided time to shape the research project into multiple phases that supported valuable opportunities for this type of research.

Opportunity 3. Community Contributions. When conducting collaborative arts-based research research, the communities you work with initiate their own opportunities. During this project, we found that communities we collaborate with were often so excited and committed to this project they would find ways to contribute towards the project in any way they could. For example, one community "received … [funds] from the… Foundation to support the technology needs for phase 2" (noted on July 28, 2017). These funds supported the purchase of televisions and laptops needed to deliver the program.

We also found that the project initiated further opportunities for communities beyond the dance program itself. For example, the research team reflected that "Alzheimer Society didn't have a support group in [this community] before this program. Starting the dance class encouraged the development of a support group" (noted on September 5, 2018). The project brought a community together, and once they did, they were further inspired to create more reasons to come together. This perhaps is one of the most meaningful opportunities that collaborative arts-based research initiatives hold.

Challenges

Arts-based research is not without its challenges. In this reflexive and dialogical analysis, we identified four key areas that presented some challenge arising from collaborative arts-based research on Sharing Dance.

Challenge 1. Multisector partnerships. Navigating partnerships from varying disciplines and sectors was challenging at times. When working with multiple partners from different backgrounds, approval processes sometimes took longer than expected because different organizations had different required processes. As noted in a team meeting discussion, "We discussed the need to streamline approval processes. The media release was more challenging than it should be and we should work to clarify [approval] processes" (noted on August 31, 2018). At times, it was also hard to align the phases of research with different partners. What worked well timing-wise for some organizations did not work for others. A team discussion noted, "…concern about starting P3 in March (one stakeholder's preferred time), mainly because it may not be a good time for … [some on the research team and community stakeholders], but we could try to accommodate" (noted on January 26, 2018). The partnerships and collaborative spirit are what allowed this project to run as well as it did. It is important to consider the types of challenges that present for projects with partners from different sectors, but with enough determination, patience, and commitment from all parties, the challenges presented in this project were overcome.

Challenge 2. Community dynamics. Community challenges related to working on collaborative arts-based research on Sharing Dance were also experienced during this project. Sustainability was a challenge that often presented in communities after the initial data collection phase. As noted at a team meeting, "[Community partner] declined participation in future terms of the program" (noted on October 3, 2018). This was despite the program running for three terms and having many engaged people in their community. The reasons provided for discontinuing the program were both related to challenges internally within the community organization that hosted the dance program and the added costs associated with maintaining the program (e.g., the cost of having an onsite facilitator). When the research project was no longer able to support cost and administration of the program, communities had more difficulty sustaining the dance program. The ability to sustain the program, however, varied across different communities. The research team noted "That there may be different needs in community versus residential settings" (noted on January 26, 2018). With internal infrastructure already in place, it may have been easier for residential settings (e.g., LTRC facilities) to sustain collaborative arts-based initiatives than community-based spaces (e.g., community centres).

Larger community organizations with multiple sites were also initially easier to recruit from than individual community settings. For example, "It was more challenging to recruit partners in Brandon that were all new individual partnerships than the experience in Peterborough… where one organization had multiple sites" (noted on April 6, 2018). These community challenges highlight how receiving and sustaining arts-based ageing programs will often be determined by factors in the community unrelated to potential demand, need, or benefit for participants.

Challenge 3. Vulnerable populations. The project experienced some challenges specific to working with people living with dementia. At times, the research team noted collecting data from people with dementia was difficult. "Some [participants] were finding all of the paperwork a barrier to participation or at least not something people enjoyed doing" (noted on September 26, 2018). As participants' dementia progressed, we also lost program and research participants from earlier phases of the study. There was "surprise…

no one from the ... group signed up. ...We lost this group because of the progression of disease and challenge to get them into the community" (noted on August 24, 2018). These challenges were not insurmountable or likely unique to this project. They are nevertheless probable challenges to consider when working on a collaborative arts-based initiative with people living with dementia.

Challenge 4. Rural dimensions. Some of the challenges that presented during this project were associated with the rural context. Resources vary from one community to another, and this may be particularly true in rural settings. Reliable internet connection was one significant challenge in a lot of our communities in both Ontario and Manitoba. Addressing internet challenges came up a lot in team meetings. For example, "The program has now been running for four weeks and there have been problems with the internet in at least one site each week" (noted on October 13, 2017). Another resource challenge was the varying human resources available in different rural communities to support this project. An onsite facilitator was required for the program to run in a community, but we noted in a team meeting that "...not all of the potential facilitators we met at the site visits met the criteria required to be a facilitator (e.g., currently leading some physical activity for older adults, basic computer/internet skills"; noted on September 7, 2017). While in some communities we found facilitators who met criteria, other communities' experiences may have been affected by the lack of access to a facilitator with optimal skills for the role. When working with rural communities, it is therefore important to consider what unique challenges may present within this context.

Promising practices for research and dance

Experience working on this project, including navigating the challenges and recognizing the opportunities noted above, highlighted four key promising practices we believe should be considered when working on collaborative arts-based initiatives.

Promising Practice 1. Experiential learning/understanding. Learning through doing plays an important role when trying to understand an arts-based innovation. Right from the beginning, we endeavoured to have everyone involved in the project experience the dance program. It was important even for the researchers to experience what they studied, and this opportunity was purposely built into time working on the project.

> [We] spoke about meeting in Toronto ... This will give [the researchers] a chance to see NBS and meet some of the people involved in this project ... [and] also give [them] a chance to participate in a *Sharing Dance* ...class.
> (noted on May 12, 2017)

We also found it helpful for stakeholders and participants to have a chance to experience the program when we presented the opportunity to participate in the project. "…Going to see the space and providing information and the [dance] demo …really helped sites understand the program/research" (noted March 20, 2019). When trying to implement an arts-based ageing initiative and research study, the offer being made to a community may be different than anything they have seen before. This may make it hard to fully appreciate why communities should consider participating in the opportunity. For everyone involved, a chance to experience the innovation may therefore be an extremely helpful practice to prioritize early on.

Promising Practice 2. Communication is essential. Effective communication should be a priority at all phases and stages of community-based research. Communicating with stakeholders regularly can help everyone stay updated and engaged. The research team noted, "Misunderstandings occur even when regular communication occurs. It is important that whenever decisions/promises are being made by any partner related to the project that they are clearly communicated to all involved" (noted Tuesday, April 3, 2018). After a misunderstanding at one point in the project, the research team determined that, "Copying partners on emails will be a good strategy to help keep everyone updated with progress being made" (noted Friday, April 6, 2018). Opportunity to communicate with community partners was also a helpful way for the project to be refined from one phase of research to another. "We discussed the need for an end of Phase 2 meeting to bring as many partners together as possible to reflect before Phase 3 begins" (noted on August 18, 2017). Additionally, "We discussed the need for a mid-term check-in with sites" (noted on September 26, 2018). Even when the program was in session, taking the time to communicate with each participating community provided valuable insight and was typically appreciated. When planning to engage in community-based initiatives, keeping the communities up-to-date with progress is critical to project success.

Promising Practice 3. Sustainability from the outset. When implementing collaborative arts-based initiatives, it is essential that sustainability is considered from the outset. While it may be tempting to focus first on implementing a model of the initiative that guarantees the research data is successfully collected, studying a program model that is well-resourced because of the research it is affiliated with will not speak to the limitations of the program once the researchers move on. The research team noted that once the data collection phase of the project was completed in a community, some sites "declined participation in future terms" (noted on October 3, 2018). Had questions about sustainability been considered throughout the project, the project could have supported the development of a more sustainable model. If sustainability is a goal for a collaborative arts-based initiative, it should be a consideration throughout the project and not just at the beginning and the end.

Promising Practice 4. Stay nimble! Each community you engage with will have its own response to implementing something new. Being able to adapt and change plans when challenges arise is an important part of implementing a successful collaborative arts-based research on Sharing Dance. Throughout this project, it was imperative that we recognize when a change to plans was needed. Meeting notes did a great job of capturing the long list of notable moments where we recognized something needed to be adjusted. "We discussed the need to change the delivery model of the program for Phase 2" (noted on October 13, 2017) and "ipads will also be distributed this coming week, but RAs have noticed that several participants seem to prefer email or handwritten diaries" (noted September 29, 2017), are two examples of moments where we recognized the need to adjust our approach. If you want to implement the optimal experience for a community, staying open to adjusting plans is imperative.

Related to the nimbleness needed to adjust throughout the project is the importance of staying reflexive. Adequate adjustments to this project would likely not have been made were it not for the designated weekly meetings the team held to collectively reflect on progress, successes, and failures throughout the project. Weekly reflection also provided opportunity to discuss methodological considerations. For example, "We discussed whether the poster created [to advertise about the upcoming session] primed participants to believe the dance program was about social inclusion because it is highlighted as a social experience" (noted on August 24, 2018). Meeting regularly to discuss the project provided weekly opportunity to collectively reflect on our work, how the project was progressing, and how it could be adjusted to optimize the experience for all.

Further team reflections in preparation for this book also helped us identify limitations of this project. For instance, the team meetings from which we based these promising practices failed to ask two fundamental questions: first, who is missing when considering these research practices? Second, what is missing from our research practices that would support more inclusion? Dance is a universal language, but we recognize the language, approach, and who we are will have all impacted who found the program and the accessibility of this research project. Future community situated arts-based research projects must include more of these conversations. In line with staying nimble and reflexive is the importance of any collaborative arts-based project to consider its limitation.

Concluding comments

We noted at the outset of this chapter how arts-based approaches to research – especially those that embrace *creative collaboration* – have become increasingly important for building knowledge about the experiences, representations, and understandings of ageing. As illustrated above, and as will be illuminated in the chapters that follow, the *Improving Social Inclusion for People with Dementia and Carers through Sharing Dance* project exemplifies the value proposition of collaborative arts-based ageing research; that the most impactful ageing research often involves creative teamwork among artists, scholars, and practitioners as well as, and most importantly, older people themselves and their carers. The lessons learned about the opportunities, challenges, and promising practices for collaborative arts-based research from our ongoing reflections as researchers is just one example that emerges from this book.

References

Amans, D. (2012). *Age and Dancing: Older People and Community Dance Practice*. Houndsmills, UK: Macmillan International Higher Education.

Armstrong, P., and Lowndes, R. (2018). *Creative Teamwork: Developing Rapid, Site-Switching Ethnography*. New York, NY: Oxford University Press.

Bartlett, R. (2015). Visualizing dementia activism: Using the arts to communicate research findings. *Qualitative Research*, *15*(6), 755–768. https://doi.org/10.1177/1468794114567493

Braun, V., and Clarke, V. (2006). Using thematic analysis in psychology. *Qualitative Research in Psychology*, *3*(2), 77–101. https://doi.org/10.1191/1478088706qp063oa

Cacari-Stone, L., Wallerstein, N., Garcia, A. P., and Minkler, M. (2014). The promise of community-based participatory research for health equity: A conceptual model for bridging evidence with policy. *American Journal of Public Health*, *104*(9), 1615–1623. https://doi.org/10.2105/ajph.2014.301961

Cornwall, A., and Jewkes, R. (1995). What is participatory research? *Social Science & Medicine*, *41*(12), 1667–1676. https://doi.org/10.1016/0277-9536(95)00127-s

Daykin, N. (2019). *Arts, Health and Well-Being: A Critical Perspective on Research, Policy and Practice*. Abingdon, UK: Routledge.

Foster, V. (2015). *Collaborative Arts-Based Research for Social Justice*. Abingdon, UK: Routledge.

Hatton, N. (2021). *Performance and Dementia: A Cultural Response to Care*. Switzerland: Springer Nature.

Herron, R., Skinner, M., Bar, R., Kontos, P., and Menec, V. (2020). Beyond therapy: Exploring the potential of dance to improve social inclusion for people with dementia. In R. Hunt and S. Atkinson (Eds.), *Geohumanities and Health*. Switzerland: Springer.

Houston, S. (2019). *Dancing with Parkinson's*. Bristol, UK: Intellect Books.

Kay, J., and Hammond, W. (2018). *The Expressive Lives of Elders Folklore, Art, and Aging*. Bloomington, IN: Indiana University Press.

Kontos, P., Grigorovich, A., Kosurko, A., Bar, R. J., Herron, R. V., Menec, V. H., and Skinner, M. W. (2021). Dancing with dementia: Exploring the embodied dimensions of creativity and social engagement. *The Gerontologist*, *61*(5), 714–723. https://doi.org/10.1093/geront/gnaa129

Kosurko, A., Herron, R., Skinner, M., Kontos, P., Menec, V., and Bar, R. (2018). *Improving social inclusion for Canadians with dementia and carers through Sharing Dance: Peterborough pilot phase three report*. Peterborough, ON: Trent University.

Kosurko, A., Herron, R., Skinner, M., Kontos, P., Menec, V., and Bar, R. (2019). *Improving social inclusion for Canadians with dementia and carers through Sharing Dance: Brandon pilot phase two report*. Peterborough, ON: Trent University.

Kosurko, A., Herron, R., Skinner, M., Kontos, P., Menec, V., and Bar, R. (2020a). *Improving social inclusion for Canadians with dementia and carers through Sharing Dance: Brandon pilot phase three report*. Peterborough, ON: Trent University.

Kosurko, A., Herron, R. V., Skinner, M. W., Bar, R. J., Kontos, P., Grigorovich, A., and Menec, V. (2020b). Opportunities and challenges of digital delivery of Sharing Dance seniors for social inclusion. *Romanian Journal of Communication and Public Relations*, *22*(2), 23–37.

Kosurko, A., Arminen, I., Herron, R., Skinner, M., and Stevanovic, M. (2021a). Observing social connectedness in a digital dance program for older adults: An EMCA approach. *International Conference on Human-Computer Interactionm*, 393–404. https://doi.org/10.1007/978-3-030-78108-8_29

Kosurko, A., Skinner, M., Herron, R. Kontos, P., Menec, V., Bar, R., and Grigorovich, A. (2021b). *Improving social inclusion for Canadians with dementia and carers through Sharing Dance: Final report*. Peterborough, ON: Trent University.

Kosurko, A., Herron, R. V., Grigorovich, A., Bar, R. J., Kontos, P., Menec, V., and Skinner, M. W. (2022). Dance wherever you are: The evolution of multimodal delivery for social inclusion of rural older adults. *Innovation in Aging*, *6*(2), iga058. https://doi.org/10.1093/geroni/igab058

Kosurko, A., Wrathall, M., Skinner, M., Herron, R., Kontos, P., Menec, V., and Bar, R. (2017). *Improving social inclusion for Canadians with dementia and carers through Sharing Dance: Peterborough pilot project planning meeting report*. Peterborough, ON: Trent Centre for Aging & Society, Trent University.

McNiff, S. (2008). 3 art-based research. In J. G. Knowles and A. L. Cole (Eds.), *Handbook of the Arts in Qualitative Research: Perspectives, Methodologies, Examples, and Issues* (pp. 29–41). SAGE Publications, Inc. https://dx.doi.org/10.4135/9781452226545.n3

Miller, E. (2021). *Creative Arts-Based Research in Aged Care: Photovoice, Photography and Poetry in Action*. Abingdon, UK: Routledge.

Minkler, M. (2005). Community-based research partnerships: Challenges and opportunities. *Journal of Urban Health*, *82*(2_suppl_2), ii3–ii12.

Mobjörk, M. (2010). Consulting versus participatory transdisciplinarity: A refined classification of transdisciplinary research. *Futures*, *42*(8), 866–873. https://doi.org/10.1016/j.futures.2010.03.003

Nguyen, T., Graham, I. D., Mrklas, K. J., Bowen, S., Cargo, M., Estabrooks, C. A., Kothari, A., Lavis, J., Macaulay, A. C., MacLeod, M., Phipps, D., Ramsden, V. R., Renfrew, M. J., Salsberg, J., and Waller, N. (2020). How does integrated knowledge translation (IKT) compare to other collaborative research approaches to generating and translating knowledge? Learning from experts in the field. *Health Research Policy and Systems*, *18*(1). https://doi.org/10.1186/s12961-020-0539-6

Novek, S., and Wilkinson, H. (2019). Safe and inclusive research practices for qualitative research involving people with dementia: A review of key issues and strategies. *Dementia*, *18*(3), 1042–1059. https://doi.org/10.1177/1471301217701274

Skinner, M. W., Herron, R. V., Bar, R. J., Kontos, P., and Menec, V. (2018). Improving social inclusion for people with dementia and carers through sharing dance: A qualitative sequential continuum of care pilot study protocol. *BMJ Open*, *8*(11), e026912. https://doi.org/10.1136/bmjopen-2018-026912

Statistics Canada. 2022. (table). Census Profile. 2021 Census of Population. Statistics Canada Catalogue No. 98-316-X2021001. Ottawa. Released April 27, 2022. https://www12.statcan.gc.ca/census-recensement/2021/dp-pd/prof/index.cfm?Lang=E (accessed May 25, 2022).

Warrener, S., Wrathall, M., Skinner, M., Herron, R., Kontos, P., Menec, V., and Bar, R. (2018a). *Improving social inclusion for Canadians with dementia and carers through Sharing Dance: Peterborough pilot phase two report*. Brandon, MB: Brandon University.

Warrener, S., Wrathall, M., Skinner, M., Herron, R., Kontos, P., Menec, V., and Bar, R. (2018b). *Improving social inclusion for Canadians with dementia and carers through Sharing Dance: Brandon pilot phase one report*. Brandon, MB: Brandon University. www.sdseniorsdresearch.com

4 Critical reflections on arts for ageing, dementia, and health

Rachel Bar and Pia Kontos

Introduction

The creative impulse is essential to the experience of being human. Indeed, in every prehistoric, ancient, and contemporary culture there is evidence of human creativity (Camic, 2008). This connection between the human condition and the arts has a significant role to play in health and well-being. So strong is this connection that it has been argued that the "birth of art was also the birth of arts in health" (Fancourt, 2017, p. 6). The evidence base linking arts engagement to health and well-being comprises research that spans a range of methodologies and practices. As we will explore in this chapter, much of this research has focused on the therapeutic and clinical benefits of engaging with the arts. While this growing body of work offers important insights regarding particular health outcomes associated with engagement with the arts, our concern is that it impoverishes understandings of the arts and perpetuates their restriction to a therapeutic end (Kontos et al., 2020). We can see this in the context of practice; with the biomedicalization and the pharmaceuticalization of ageing and dementia, the arts have primarily been implemented as a non-pharmacological means to generate targeted health outcomes. This is despite that engagement with the arts is increasingly advocated to promote, preserve, and sustain human flourishing (Dupuis et al., 2016).

Human flourishing has been conceptualized as an optimal state since the time of Aristotle (Oberholzer, 2019). Seligman (2011) includes positive relationships, accomplishments, meaning and positive emotion as key components of flourishing. Physical, mental, and social health are not themselves determinants of human flourishing (Oberholzer, 2019) and thus even in the face if disability, individuals still have the potential to flourish. Oberholzer (2019) makes this argument for children with disabilities and Kontos and Grigorovich (2018a) make this argument for people living with dementia. As an artist and social researchers working in the health research field, we are committed to the interactive, educational, and emancipatory power of the arts to redress inequities and to foster a more inclusive and life-enriching society wherein human flourishing is the guiding ethic for older adults. However, in our respective research programs and practices, we find

DOI: 10.4324/9781003260691-4

ourselves often having to work 'against the grain,' specifically to resist barriers to achieving the transformative potential of the arts within a health context (e.g., long-term care sector; clinical populations). These include but are not limited to the interventionist logic and associated reductionist methodologies that dominate health research and even arts-based research (Parsons et al., 2017), and also the art-for-therapy culture (Kontos and Grigorovich, 2018a, 2018b).

Here we wish to offer some reflections on the ways that Sharing Dance Older Adults challenges the resulting inequities that persist, the tensions we have encountered in doing this work and how we addressed them, and what we have learned from Sharing Dance that can inform the broader movement to support human flourishing. To contextualize our reflections, we begin with a brief review of an expanding body of evidence that supports the positive correlation between engagement with the arts and ageing, health, and well-being.

What does the evidence tell us? Intersections between the arts, ageing, health, and well-being

Many benefits of engaging with the arts are intuitive, such as the stimulation of the imagination and reflection, facilitating access to a range of emotions, and yielding opportunities to make meaningful contributions to the self and others. Additionally, there is an expanding body of evidence from research to support the contention that the arts have an important contribution to make to ageing, health, and well-being. It goes beyond the scope of this chapter to provide an extensive review of this research. However, given our focus on ageing and dementia, we provide here a brief overview of some of the health benefits of engagement with the arts in older adulthood, recovery from illness, and management of long-term conditions. This overview was informed by two comprehensive reports on arts for health and well-being (Gordon-Nesbitt, 2017; WHO, 2019).

Opportunities to engage with the arts in older adulthood have been found to support a range of health indicators. Singing activities have led to higher overall ratings of physical health, fewer doctor visits, less medication use, and fewer falls than the usual activity (Cohen et al., 2006). Older people dancing has been found to support and improve balance, strength, and posture to name just a few physical benefits, and simultaneously addresses social isolation (Connolly and Redding, 2010). Older adults who participate in cultural outings (e.g., museum, theatre, and cinema) have been found to be at a reduced risk of depression, even when controlling for health and social confounds (Fancourt and Tymoszuk, 2019), and generally experience a slower rate of cognitive decline (Fancourt et al., 2018).

In relation to recovery from illness in older adults, much of the research has explored the impact of engaging with the arts in stroke rehabilitation. Listening to music has been found to stimulate regions of the brain responsible for attention, motor function, memory, and emotional processing (Särkämö and Soto, 2012; Särkämö et al., 2014). These benefits have been accompanied by improvements in the recovery of verbal memory, executive function, reductions in confusion and depression, and improved mental health and well-being (Baylan et al., 2016; Fujioka et al., 2018; Raglio et al., 2017). For motor rehabilitation after stroke, music therapy and dance have been found to improve upper- and lower-limb motor function, muscular weakness, gait, balance, and stride length (Patterson et al., 2018; Zhang et al., 2016). Engagement with drawing and painting has similarly been found to stimulate physical and cognitive abilities following a stroke while also increasing confidence, self-determination, and quality of life (Beesley et al., 2011). In the context of the management of long-term conditions, Parkinson's disease and the arts have been the focus of much research including music and dance, which has been found to improve balance and functional mobility (dos Santos Delabary et al., 2017), and singing can help to reduce the symptoms of a weak voice, and reduce imprecise articulation or impaired stress or rhythm in speech (Barnish et al., 2016; Di Benedetto et al., 2009; Han et al., 2018).

In response to concerns about the overreliance on pharmacotherapies in dementia care, non-pharmacological interventions are now recommended before resorting to psychotropic medication. It is with the movement to non-pharmacological interventions that arts-based programs have become cornerstones of dementia care, including dance/movement, music, drama, visual arts, and combination art therapies. Music therapy has the largest evidence base to support its efficacy (Moreno-Morales et al., 2020), which primarily focuses on the application of music as a 'treatment' to ameliorate emotional, behavioural, social or cognitive 'problems' (Chang et al., 2015; van der Steen et al., 2017). Visual arts, such as drawing and painting, are the next most common arts-based modality in the literature and are discussed in terms of application in diagnosis and assessment, promoting non-verbal communication, enhancing opportunities for reminiscence (Beard, 2012). More recently, dance is receiving increasing attention in dementia care for its beneficial effects for people living with dementia. Dance is used as a psychotherapeutic and rehabilitative treatment modality by combining the physical benefits of exercise with psychosocial therapeutic benefits (Earhart, 2009; Karkou and Meekums, 2017). Dance therapists interweave dance and psychotherapy based on the premise that movement reflects patterns of cognitive reflection and emotion and that remediating posture or autonomic nervous system activity will influence neurological processes thought to be implicated in mood, perception, sensory experience, cognition, and well-being (Guzmán-García et al., 2013; Ho et al., 2015).

This collective evidence base spans a wide range of methodologies. However, it is mostly concentrated within clinical research where 'impact' is viewed as outcome-based, concrete, and easily measured, and thus quantitative measures prevail. Qualitative methods including case studies, interviews, focus groups, and observations are slowly gaining acceptance in arts-based research with calls for a more nuanced and expansive approach to capturing the impact of engaging with the arts (Parsons et al., 2017). Empowerment, communication and meaningful self-expression, communal spirit, pleasurable experience, and sociability are just some of the benefits of engagement with the arts that have been identified with qualitative methods (de Medeiros and Basting, 2013; Kontos et al., 2020). Yet, these are typically considered side benefits and not the primary intention of implementing arts-based programs. Deep-seated assumptions about science no doubt are influencing this perception; with the continued dominance of medical research criteria wherein controlled trials and quantitative outcomes measures are held as the gold standard, qualitative assessments and the kind of impact they capture are often viewed sceptically. Scientific efforts in turn influence political attention (e.g., funding) and the organization and delivery of programming, and thus the continued focus on biomedical constructions/methods such as pathology, clinical trials, and pharmacological treatments reproduces in practice their instrumental reduction to the application as a therapeutic tool.

We offer in the next section some reflections on the ways that Sharing Dance challenges the inequities that result from this reductionism, why this is so important, and how we can draw on the learnings from the development and evaluation of Sharing Dance to inform a broader movement to support the arts for human flourishing.

A conversation

Sharing Dance's approach to ageing, dementia, and health

PIA KONTOS: I came to learn about Sharing Dance when I was invited to participate on a grant application to the Canadian Institutes of Health Research (CIHR) to study Sharing Dance. That is when I first met you. I was immediately intrigued by the invitation since creativity is something that I had observed in so much of the research I conducted with people living with dementia in both long-term residential care (LTRC) and community care settings. For example, in an ethnographic study I conducted in a LTRC home, I would often see residents singing or dancing spontaneously in the hallways, living room, or dining room. The following is an interaction I remember well (Kontos and Grigorovich, 2018c, p. 170):

> A personal support worker came into the living room and turned on the radio. She paused to listen to the soft music and then walked over to Abe who was seated on the sofa, holding her hands out for him to take. He grasped her hands and stood up. Standing close together and facing each other, they naturally moved into dancing position: he put one hand on her waist and she placed one hand on his shoulder, and their free hands clasped together and were held out to the side. Once they were in position the two of them began to move, slowly turning in unison, following the gentle pace of the music. When the song was over, Abe released his hands, stepped back from the personal support worker slightly and bowed. She smiled and gave him a hug and then helped him back onto the sofa.

With an interest in exploring creativity in a more focused way, I led a study on elder-clowning (Kontos et al., 2017) where I found that often with only embodied self-expression and/or nonsensical speech, residents would initiate, modify, and co-construct exquisite moments of engagement through dance, song, humour, and improvisational play. Examples of interactions that I observed between the residents of a LTRC home and the elder-clowns included: song and music, such as singing with residents their favourite songs with the accompaniment of a miniature ukulele, or co-constructing with them improvised songs; witty, playful scenarios involving, for example, teasing the elder-clowns by playfully pretend-kicking their buttocks as clowns bent over, to which the elder-clowns responded with exaggerated pratfalls, sound explosions, and facial animations; supporting sadness with soft reassuring touch rather than trying to change the emotional timbre; and artistic expression by residents through the elder-clowns' provision of pens and sketch pads, or more imaginative engagement of residents such as an elder-clown creatively miming an artist painting a canvas. Yet, with more research on arts-based innovations in dementia care, I came to learn that creative self-expression was most commonly valued as a therapeutic rather than being central to life enrichment. For example, there was a woman who loved to sing during her meals, and the staff would always ask her to stop singing and to finish her meal so that they could clear and then set the table for the next meal. It was always difficult to observe this since her singing brought her and the other residents in the dining room such delight. There was also an absurdity to it given that just after the meal she would be taken to music therapy. These experiences of mine left me feeling deeply troubled about what this denies for people living with dementia, and it also triggered in me a deep commitment to ensuring that people living with dementia are given equal opportunities to participate in life – including creative pursuits.

Given these academic and activist interests of mine, I loved the idea of supporting dance in dementia care settings. However, without yet knowing anything about Sharing Dance, I was concerned that this was just like so many other arts-based programs for people living with dementia that are implemented strictly for therapeutic purposes as a non-pharmacological means to impact 'problem' behaviours and functioning. As I had discovered in a review I had recently published, dance was no exception to this. This review traces the therapeutic focus of dance to the emergence of the academic discipline of 'dance science' within the context of physical education and rehabilitation science (Kontos and Grigorovich, 2018c). As dance developed into an academic discipline, it appropriated the curricular visions of these other health sciences creating hybrid amalgams such as 'dance kinesiology' that focus on biomechanics, physiology, and motor action and control. The appeal of dance science is that it offers objective and quantitative methodologies that enhance the scientific legitimacy of dance practice. Yet, the focus becomes 'movement analyses' or 'movement profiles' which, in turn, inform dance/movement therapy interventions aimed at improving 'performance and productivity'; this elides the very art of dance and its power to support non-verbal communication and affect, sentient and tacit forms of expression, and the empowerment and pleasure that people living with dementia derive from it (Kontos and Grigorovich, 2018a). Given how fundamental these aspects are to being human, to human interaction, and to everyday existence, it seems egregious to restrict the arts to their instrumental application as a therapeutic tool to improve 'behaviours' and cognitive functioning.

I remember sharing this critique with the research team and having such a rich and engaging discussion about the focus of Sharing Dance, which I came to understand as one that fundamentally counters the art-for-therapy culture.

RACHEL BAR: I recall those discussions well; your concerns really resonated with me. I came to work on Sharing Dance wearing both the hat of an artist and researcher. As a dancer, having experienced the physical, emotional, and spiritual joy of dance for most of my life, the opportunity to help create a dance program for older people who otherwise were not able to access traditional dance experiences excited and inspired me. I was hopeful that by finding ways to share dance with older people they too would experience this joy. As a researcher, however, I initially came to the Sharing Dance project with some scepticism that we could truly capture the benefits of dance with the types of research methods I had been trained to use, which were quantitative measures. I had read dozens of research studies about the benefits of dance for older adults and felt that the quantitative nature of this work missed why it is that I believe all humans can benefit from dance. This is not to suggest that improved balance, posture, and even mood are not important benefits of dance; I just do not believe they capture how and why dance is so transformative. When I learned about the qualitative approach that was proposed for the project, I was hopeful that we could capture more of the unique qualities of dance beyond therapy that have made the art form the adaptive and universal part of the human experience I know and love.

From the very beginning, Sharing Dance was about making dance accessible. Our partners at Baycrest provided important insights and considerations regarding the kinds of physical and cognitive challenges older people may experience, but our goal was never to 'treat' these issues. Instead we wanted to offer opportunities for older people to engage in dance for the love of dance and the benefits that derive from dance for life enrichment. And so, when we began to develop the protocol for these dance classes, we looked to the structure of the other types of dance classes we offered as dance teachers to see what elements of the art form were essential. As an example, as in many traditional dance classes, Sharing Dance classes end with a réverénce. Réverénce is an opportunity at the end of a dance class to acknowledge our gratitude for the experience and traditionally acknowledges the teacher, musician, and others involved in the dance experience. Adapting this traditional dance practice into the Sharing Dance classes provided our dancers, teachers, musicians, and other participants a chance to similarly experience a calming expression of gratitude through movement.

There are several other components of the program that also reflect our attempt to create an artistic experience for our dancers. Eye-line or eye focus is a nuanced but critical detail when performing a dance, especially when there is a story or narrative to express through the dance. This is one technical element of dance that is brought out a lot in Sharing Dance. Whether following your hand as it moves or finding someone else in the room to make eye contact with as you dance, it was important that Sharing Dance included qualitative details of dance experiences that are used to make connections between our movements and to others, and to hold the art form at the centre of the innovation.

Creating movement within a creative context is another key component to the approach used in Sharing Dance. Firstly, within the context of the classes, participants are always referred to as *dancers*, which recognizes everyone as artists and thus invites participants to approach the movements in a more playful and exploratory way. This identity thereby sets the stage for the dancers to express themselves creatively. A creative cue, such as "reach for the stars" will be embraced differently by each dancer and this is exactly the point. There is no right or wrong way to express oneself in an artistic context. It therefore provides the dancers with ample opportunity to succeed in any way that they express themselves.

PIA KONTOS: These are all such important ways that Sharing Dance supports social inclusion; you make dance accessible to older adults with a range of physical and cognitive abilities, including people living with dementia by emphasizing the importance of *creative self-expression* and prioritizing expressive capacities and social interaction. This was why I enthusiastically joined the team as I saw this as an exciting opportunity to research this novel arts-based innovation, and to thereby contribute to the dearth of research on the ways that dance can support life enrichment. Shifting the focus of research on dance from therapeutic and clinical benefits to the ways that dance supports embodied, relational and affective dimensions of experience, would also help to challenge stigma and to cultivate a new culture of dementia care wherein people living with dementia are given equal opportunities to be creative to the fullest extent possible. This is an important goal of mine.

RACHEL BAR: I share your commitment. But achieving this is certainly not without tensions.

Exploring tensions in arts-based research

RACHEL BAR: I think that one of the greatest tensions involved in developing and researching Sharing Dance boiled down to a difference between art and science. Arts and sciences both share creativity as an essential component, but given the dominance of positivism (e.g., precision, quantification, detached observation), the sciences are less interested in processes of creativity and more on specific outcomes of it. In contrast to this, for artists, the purpose of the arts is dialogical, a reflective conversation and exploration between the human condition and human potential. Therefore, one of the key differences between the arts and how the sciences have engaged with them for me relates to accountability. While artists explore their craft freely, scientists by practice must define their purpose. In the health sciences in particular, accountability is essential as error may have significant deleterious effects on human life. So it felt important that Sharing Dance demonstrate accountability by clearly defining the program's approach and protocol (Tafler et al., under review). While as artists we wanted to create an artistic experience for our dancers, we also needed to create an approach with enough accountability and replicability that the health sector could support uptake by older adults.

PIA KONTOS: I too have experienced those tensions as a critical qualitative researcher in health research practice and evaluation, and as someone who is also committed to the liberatory power of the arts. In scientific knowledge production, predictable, measurable, and commercially productive research findings are more valued. This has had enormous consequences for all health researchers, but especially for critical qualitative health researchers whose political and epistemological orientation (i.e., critical hermeneutic) and research processes are in opposition to such ideologies and instrumental aims (Kontos and Grigorovich, 2018d). For example, qualitative research is accorded less funding and resources than quantitative research within academic medical/health sciences faculties, research institutes, and professional and political sectors. And even within a mixed methods design, qualitative research occupies a marginal position as it is valued primarily for improving quantitative measurement efforts or outcomes of biomedical interventions and for humanizing statistical results.

RACHEL BAR: There is also the dominance of the interventionist paradigm and the value accorded to demonstrating therapeutic and clinical benefits of arts-based innovations. All of this presented challenges for us in terms of how to pitch our study to CIHR.

Resistance strategies

RACHEL BAR: I recall our early discussions about whether to refer to Sharing Dance as an 'intervention,' and how to manage the deep-seated assumptions about 'impact' in the health sciences and the anticipated expectations of reviewers that our approach is 'outcome based' (e.g., preventing falling, improving balance, cognition, decrease 'problematic behaviours').

PIA KONTOS: I remember we really struggled to resist the pressure to massage our research into prescribed forms and formulas and to stay true to our critical qualitative approach to research (i.e., exploring complex, contextualized understandings and interrogating the status quo and who and what is served by these constructions/approaches and exposing their implication). Given the dominance of the interventionist paradigm and the misconceptions about qualitative inquiry, it was difficult to craft a compelling proposal to explore first-person lived experiences of dance and the multi-level processes as they relate to the dynamics of social inclusion.

RACHEL BAR: It was helpful to draw on scholarship that enlists the arts to address policy challenges such as mitigating social isolation and loneliness, that argues for more equitable access to the arts for older adults (including people living with dementia), and how this can strengthen local services, and promoting more cohesive communities (Kontos et al., 2021).

PIA KONTOS: We also very powerfully articulated the imperative to challenge the profound stigma that continues to be associated with dementia, the ways that engagement with arts have been able to reduce stigma, and the necessity for a qualitative approach given that reductionist methodologies are ill-equipped to capture the first-person lived experience of the phenomenon of dance as well as multi-level processes related to social inclusion (see Herron, Novek and Menec, Chapter 5). This all made for a very compelling case for what we were proposing.

Where does this success lead us?

RACHEL BAR: I do hope that our success in securing the CIHR grant and this project will help to dispel misconceptions about qualitative research in arts and health, and to inspire other researchers to challenge the instrumental reduction of dance, and the arts more generally, to their application as a therapeutic tool with quantifiable outcomes reductionism. I also hope that Sharing Dance and the research we have done helps to create a vibrant community of practice for fellow artists to nurture creativity in the context of everyday life for older adults. My experiences dancing with older adults through Sharing Dance continuously reconstitutes my conception of the richness of human capacity. Let this be everyone's experience. Finally, for both the artist and researcher interested in the arts for ageing, dementia, and life enrichment, I hope to see more of the mutual respect I experienced working on this Sharing Dance project in which the multiple ways of knowing and appreciating the arts for older adults were approached with trust, genuine curiosity, and plenty of dancing. What has Sharing Dance inspired for you?

PIA KONTOS: I see the enormous potential of Sharing Dance to model how other arts-based innovations can support the relational, interpersonal, and affective dimensions of human experience. This inspires me to engage in broad community awareness building and education to deepen learning and engagement about the capacity of older adults to be creative (well beyond for example the typical program offerings for people living with dementia such as 'paint-by-number') and the infinite value of the arts in supporting human flourishing. There is much work to be done to foster the development of dance lives, which is what Sharing Dance supports, to learn from this form of experience, and to critically interrogate barriers to it being realized in practice. I hope that our learnings reflected on here will be taken up by other researchers, artists, policy-makers, and activists equally committed to ensuring that older adults have the opportunities to participate in the arts to the fullest extent possible.

Concluding comments

In our reflections, we highlight several important ways in which Sharing Dance and our approach to studying this arts-based innovation prioritize human flourishing. Our commitment to this approach was not without its tensions; however, utilizing qualitative methods helped us to explore processes of engagement and more nuanced, fluid, and expansive understandings of impact, which was so critical for capturing the ways in which dance as an art form can support human flourishing. As Parsons et al. (2017, p. 169) have argued, impact is "aesthetically mediated, embodied, and engaging" and to this end, the complex and multifaceted nature of arts-based health research is at the core of its methodological power (see Skinner and Bar, Chapter 3).

There are other examples of innovative dance programs like the Mark Morris Dance Group's internationally renowned Dance for PD® program that fits broadly within a community dance movement (see Bar and Dalrymple, Chapter 2) that sees dance artists increasingly creating opportunities for diverse communities to engage in dance. There are also arts-based innovations to support life enrichment for people living with dementia such as TimeSlips, the BUDI Symphony project (Bournemouth University Dementia Institute, 2015), Meet me at the MOMA, and the Bitove Method. However, more work is needed to challenge the reductionism of the art-for-therapy culture that continues to dominate the health field and to strengthen the critical emerging discourse on the arts as a proponent of human flourishing for everyone, including individuals who live with cognitive and/or physical disabilities. Key to this work will be a multifaceted approach that entails cross-disciplinary, cross-sectoral, and creative collaborations of the nature that is advocated for by the contributing authors of the collective chapters in this book. This will importantly include partnerships between researchers from diverse disciplines, artists, older adults, carers, and community organizations. This transdisciplinary approach to arts-based research resists siloed boundaries of disparate disciplines and their respective assumptions, methods, and frames of reference through greater participatory collaboration across disciplines and sectors. Such an approach would help to unsettle deeply entrenched assumptions within the academy about what counts as research, as evidence, as legitimate inquiry. It rejects a hierarchy of knowledge and valourizes different ways of knowing that exist outside of academic frameworks and may be more relevant to artistic and community stakeholders.

Such an approach would also help to challenge the instrumental reduction of the arts to therapy, and support a new ethical standard for nurturing creativity in older adults in everyday life. The cultivation of the capacity for such engagement can contribute to an individual's flourishing, and yet the ethical significance of creativity in the lives of older adults is far from realized (Kontos et al., 2020). It is our hope that our reflections here and the collective insights of the other chapters of this book equip scholars, practitioners, and experiential stakeholders to resist dominant discourses/approaches, to value "beauty and artistry, movement, rhythm, color and texture in everyday life" (Denzin, 2003, p. 113), and to support engagement with the arts to the fullest extent possible for the human good.

References

Barnish, J., Atkinson, R. A., Barran, S. M., and Barnish, M. S. (2016). Potential benefit of singing for people with Parkinson's disease: A systematic review. *Journal of Parkinson's Disease*, *6*(3), 473–84. https://doi.org/10.1136/jech-2017-ssmabstracts.104

Baylan, S., Swann-Price, R., Peryer, G., and Quinn, T. (2016). The effects of music listening interventions on cognition and mood post-stroke: A systematic review. *Expert Review of Neurotherapeutics*, *16*(11), 1241–1249. https://doi.org/10.1080/14737175.2016.1227241

Beard, R. L. (2012). Art therapies and dementia care: A systematic review. *Dementia*, *11*(5), 633–656. https://doi.org/10.1177/1471301211421090

Beesley, K., White, J. H., Alston, M. K., Sweetapple, A. L., and Pollack, M. (2011). Art after stroke: The qualitative experience of community dwelling stroke survivors in a group art programme. *Disability and Rehabilitation*, *33*(23–24), 2346–2355. https://doi.org/10.3109/09638288.2011.571333

Bournemouth University Dementia Institute. (2015). BUDI Orchestra summary. Retrieved May 31, 2016, from http://blogs.bournemouth.ac.uk/dementia-institute/files/2015/02/BUDI-Orchestra-Summary1.pdf

Camic, P. M. (2008). Playing in the mud: Health psychology, the arts and creative approaches to healthcare. *Journal of Health Psychology*, *13*, 289. https://doi.org/10.1177/1359105307086698

Chang, Y.-S., Chu, H., Yang, C-Y., Tsai, J-C., Chung, M-H., Liao, Y-M., Chi, M-J., Liu, M. F., and Chou, K.-R. (2015). The efficacy of music therapy for people with dementia: A meta-analysis of randomised controlled trials. *Journal of Clinical Nursing*, *24*(23–24), 3425–3440. https://doi.org/10.1111/jocn.12976

Cohen, G. D., Perlstein, S., Chapline, J., Kelly, J., Firth, K. M., and Simmens, S. (2006). The impact of professionally conducted cultural programs on the physical health, mental health, and social functioning of older adults. *The Gerontologist*, *46*(6), 726–734. https://doi.org/10.1093/geront/46.6.726

Connolly, M. K., and Redding, E. (2010). *Dancing towards well-being in the third age: Literature review on the impact of dance on health and well-being among older people*. London: Trinity Laban Conservatoire of Music and Dance. See Dance for Lifelong Wellbeing at the Royal Academy of Dance. https://www.artshealthresources.org.uk/docs/dancing-towards-wellbeing-in-the-third-age-literature-review-on-the-impact-of-dance-on-health-and-well-being-among-older-people/

Denzin, N. K. (2003). *Performance Ethnography: Critical Pedagogy and the Politics of Culture*. Thousand Oaks, CA: Sage.

Di Benedetto P., Cavazzon, M., Mondolo, F., Rugio, G., Peratoner, A., and Biasutti, E. (2009). Voice and choral singing treatment: A new approach for speech and voice disorders in Parkinson's disease. *European Journal of Physical Rehabilitation Medicine*, *45*(1), 13–19.

dos Santos Delabary, M., Komeroski, I. G., Monteiro, E. P., Costa, R. R., and Haas, A. N. (2017). Effects of dance practice on functional mobility, motor symptoms and quality of life in people with Parkinson's disease: A systematic review with meta-analysis. *Aging clinical and experimental research*, *30*(7), 727–735.

Dupuis, S. L., Kontos, P., Mitchell, G., Jonas-Simpson, C., and Gray, J. (2016). Re-claiming citizenship through the arts. *Dementia: The International Journal of Social Research and Practice (Special Issue: Citizenship and Dementia)*, *15*(3), 358–380. https://doi.org/10.1177/1471301216637206

Earhart, G. M. (2009). Dance as therapy for individuals with Parkinson disease. *European Journal of Physical and Rehabilitation Medicine*, *45*(2), 231.

Fancourt, D. (2017). *Arts in Health: Designing and Researching Interventions*. Oxford: Oxford University Press.

Fancourt, D., and Tymoszuk, U. (2019). Cultural engagement and incident depression in older adults: Evidence from the English Longitudinal Study of Ageing. *The British Journal of Psychiatry*, *214*(4), 225–229. https://doi.org/10.1192/bjp.2018.267

Fancourt, D., Steptoe, A., and Cadar, D. (2018). Cultural engagement predicts changes in cognitive function in older adults over a 10 year period: Findings from the English Longitudinal Study of Ageing. *Scientific Reports*, *8*(1), 10226. https://doi.org/10.1038/s41598-018-28591-8

Fujioka, T., Dawson, D. R., Wright, R., Honjo, K., Chen, J. L., and Chen, J. J. (2018). The effects of music supported therapy on motor, cognitive, and psychosocial functions in chronic stroke. *Annals of the New York Academy of Sciences*, *1423*(1), 264–274. https://doi.org/10.1111/nyas.13706

Gordon-Nesbitt, R. (2017). *Creative Health: The Arts for Health and Wellbeing*. All-Party Parliamentary Group on Arts, Health and Wellbeing. http://www.artshealthandwellbeing.org.uk/appg-inquiry/Publications/Creative_Health_Inquiry_Report_2017_-_Second_Edition.pdf

Guzmán-García, A. H. J. C., Hughes, J. C., James, I. A., and Rochester, L. (2013). Dancing as a psychosocial intervention in care homes: A systematic review of the literature. *International Journal of Geriatric Psychiatry*, *28*(9), 914–924. https://doi.org/10.1002/gps.3913

Han, E. Y., Yun, J. Y., Chong, H. J., and Choi, K. G. (2018). Individual therapeutic singing program for vocal quality and depression in Parkinson's disease. *Journal of Movement Disorders*, *11*(3), 121–128. https://doi.org/10.14802/jmd.17078

Ho, R. T. H., Cheung, J. K. K., Chan, W. C., Cheung, I. K. M., and Lam, L. C. W. (2015) A 3-arm randomized controlled trial on the effects of dance movement intervention and exercises on elderly with early dementia. *BMC Geriatrics*, *15*, 127.

Karkou, V., and Meekums, B. (2017). Dance movement therapy for dementia. *Cochrane Database of Systematic Reviews*, (2), 1–29. https://doi.org/10.1186/s12877-015-0123-z

Kontos, P., and Grigorovich, A. (2018a). Integrating citizenship, embodiment, and relationality: Towards a reconceptualization of dance and dementia in long-term care. *Journal of Law, Medicine & Ethics (Special Issue: Perspectives on Alzheimer's Disease: Ethical, Legal, and Social Issues)*, *46*(3), 717–723. https://doi.org/10.1177/1073110518804233

Kontos, P., and Grigorovich, A. (2018b). Rethinking musicality in dementia as embodied and relational. *Journal of Aging Studies (Special Issue: Ageing Body and*

Society – Critical Perspectives, Future Challenges), *45*, 39–48. https://doi.org/10.1016/j.jaging.2018.01.006

Kontos, P., and Grigorovich, A. (2018c). Dancing with dementia: Citizenship, embodiment, and everyday life in long-term care. In S. Katz (Ed.), *Ageing and Everyday Life: Materialities and Embodiments* (pp.163–179). Bristol: Policy Press.

Kontos, P., and Grigorovich, A. (2018d). 'Sleight of hand' or 'selling our soul'? Surviving as critical qualitative health researchers in a positivist world. *Forum Qualitative Sozialforschung / Forum: QualitativeSocial Research*, *19*(2), Art. 25.

Kontos, P., Miller, K. L., Mitchell, G., and Stirling-Twist, J. (2017). Presence redefined: The reciprocal nature of engagement between elder-clowns and persons with dementia. *Dementia: The International Journal of Social Research and Practice*, *16*(1), 46–66. https://doi.org/10.1177/1471301215580895

Kontos, P., Grigorovich, A., and Colobong, R. (2020). Towards a critical understanding of creativity and dementia: New directions for practice change. *International Practice Development Journal (Special Issue: Critical perspectives on person, care and aging)*, *10*(3). https://doi.org/10.19043/ipdj.10suppl.003

Kontos, P., Radnofsky, M., Fehr, P., Belleville, M. R., Bottenberg, F., Fridley, M., Massad, S., Grigorovich, A., Carson, J., Rogenski, K., Carpenter, K.S., Dupuis, S., Battalen, J., McDonagh, D., Fassbender, K., and Whitehouse, P. (2021). Separate and unequal: A time to reimagine dementia. *Journal of Alzheimer's Disease*, *80*(4), 1395–1399. https://doi.org/10.3233/jad-210057

de Medeiros, K., and Basting, A. (2013). "Shall I compare thee to a dose of Donepezil?": Cultural arts interventions in dementia care research. *The Gerontologist*, *54*(3), 344–353. https://doi.org/10.1093/geront/gnt055

Moreno-Morales, C., Calero, R., Moreno-Morales, P., and Pintado, C. (2020). Music therapy in the treatment of dementia: A systematic review and meta-analysis. *Frontiers in Medicine*, *7*, 160. https://doi.org/10.3389/fmed.2020.00160

Oberholzer, A. E. (2019). Human flourishing in the lives of children living with a disability. *Journal of Disability & Religion*, *23*(2), 127–148. https://doi.org/10.1080/23312521.2019.1571468

Parsons, J. A., Gladstone, B. M., Gray, J., and Kontos, P. (2017). Re-conceptualizing 'impact' in art-based health research. *Journal of Applied Arts & Health*, *8*(2), 155–173. https://doi.org/10.1386/jaah.8.2.155_1

Patterson, K. K., Wong, J. S., Nguyen, T. U., and Brooks, D. (2018). A dance program to improve gait and balance in individuals with chronic stroke: A feasibility study. *Topics in Stroke Rehabilitation*, *25*(6), 410–416. https://doi.org/10.1080/10749357.2018.1469714

Raglio, A., Zaliani, A., Baiard, I. P., Bossi, D., Sguazzin, C., Capodaglio, E., Imbriani, C., Gontero, G., and Imbriani, M. (2017). Active music therapy approach for stroke patients in the post-acute rehabilitation. *Neurological Sciences*, *38*(5), 893–897. https://doi.org/10.1007/s10072-017-2827-7

Särkämö, T., and Soto, D. (2012). Music listening after stroke: Beneficial effects and potential neural mechanisms. *Annals of the New York Academy of Sciences*, *1252*(1), 266–81. https://doi.org/10.1111/j.1749-6632.2011.06405.x

Särkämö, T., Ripollés, P., Vepsäläinen, H., Autti, T., Silvenno, H. M., Salli, E., Laitinen, S., Forsblom, A., Soinila, S., and Rodriguez-Fornells, A. (2014). Structural changes induced by daily music listening in the recovering brain after middle cerebral artery stroke: A voxel-based morphometry study. *Frontiers in Human Neuroscience*, *8*, 245. https://doi.org/10.3389/fnhum.2014.00245

Seligman, M. E. (2011). Flourish: A visionary new understanding of happiness and well-being. *Policy*, *27*(3), 60–61.

van der Steen, J. T., van Soest-Poortvliet, M. C., van der Wouden, J. C., Bruinsma, M. S., Scholten, R. J. P. M., and Vink, A. C. (2017). Music-based therapeutic interventions for people with dementia. *The Cochrane Library*. http://onlinelibrary.wiley.com/doi/10.1002/14651858.CD003477.pub3/abstract

Tafler, M., Bar, R. J., Kim, S., and Sodums, D. (under review). The development of Sharing Dance Older Adults: Baycrest and Canada's National Ballet School make dance accessible to older adults with dementia. Manuscript available from corresponding author: rbar@nbs-enb.ca.

World Health Organization. (2019). *What is the evidence on the role of the arts in improving health and well-being? A scoping review*. World Health Organization. Regional Office for Europe. https://apps.who.int/iris/bitstream/handle/10665/329834/9789289054553-eng.pdf

Zhang, Y., Cai, J., Zhang, Y., Ren, T., Zhao, M., and Zhao, Q. (2016). Improvement in stroke-induced motor dysfunction by music-supported therapy: A systematic review and meta-analysis. *Scientific Reports Rep*, *6*, 38521. https://doi.org/10.1038/srep38521

5 Advancing age- and dementia-related social inclusion through *Sharing Dance*

Rachel Herron, Sheila Novek, and Verena Menec

Introduction

Older adults have a right to meaningful participation in activities and relationships within their environments, yet some older people experience barriers to achieving this basic goal. Social attitudes, policies, and practices can prevent older adults, especially those living with dementia from participating fully in the social world (Shakespeare et al., 2019; Swaffer, 2015). For example, when ageist attitudes devalue older adults and discriminate against people living with disabilities, they can influence how older adults see themselves (i.e., self-stigma) as well as what activities and services they feel comfortable engaging with (i.e., label avoidance). Furthermore, ageism can influence policies and practices within communities, institutions, and countries (WHO, 2021). These examples underscore the multifaceted challenges of addressing social exclusion and promoting social inclusion, which can sometimes be experienced alongside one another as older adults negotiate activities and relationships within their environment (Bartlett, 2007).

Social connection has long been understood as important to the health and well-being of older adults. However, the more recent interest in social inclusion as a means of health promotion reflects a shift within health and ageing research from focusing on ageing bodies and medical interventions to social models of health (Moody and Phinney, 2012). Over the past decade, the field of dementia studies has undergone a similar shift from a biomedical emphasis towards understanding the experiences of people living with dementia within their social, environmental, and political contexts (Bartlett and O'Connor, 2010; Pinkert et al., 2019). This emphasis on social and community care contexts has generated an increasing focus on related (and often conflated) concepts such as social isolation, social connection, and social inclusion. In light of the large literature on the negative health impacts of social isolation and loneliness (e.g., Courtin and Knapp, 2017; Holt-Lunstad et al., 2015; Leigh-Hunt et al., 2017), promoting social inclusion has come to be seen as a health solution.

The social inclusion of older adults and people living with dementia is a key tenet of global action plans on ageing and dementia as well as national legislation and policies. In 2002, the United Nation's foundational

DOI: 10.4324/9781003260691-5

Madrid International Plan of Action on Ageing called for the "full inclusion and participation of older persons in societies" (p. 10). Since then, social inclusion has also been promoted by the global age-friendly communities movement (WHO, 2007). The WHO Age-Friendly Cities framework conceptualizes respect and social inclusion as a core domain of age-friendly environments. In more recent years, the social inclusion of people living with dementia has been increasingly adopted as a goal of global strategies, public policy, and service provision (WHO, 2012). The WHO's *Global Action plan on the Public Health Response to Dementia* (2017), for example, calls for the development of age- and dementia-friendly environments that enable participation and inclusion. Like the age-friendly communities concept, dementia-friendly communities refer to supportive and inclusive environments that enable people living with dementia to exercise choice and to participate in their social networks and communities (Hebert and Scales, 2019; Hung et al., 2020; Wiersma and Denton, 2016). These influential global strategies frame social inclusion as a right, an outcome shaped by social and material environments, and a strategy to remove barriers and promote quality of life.

In higher income countries such as Canada, national dementia strategies identify inclusion as a right and call for the development of dementia-inclusive communities. The inclusion of individuals with dementia is listed as a tenet of the guiding principle, "respect human rights", one of the five principles directing action on dementia (Public Health Agency of Canada, 2019). The strategy also promotes the development of dementia-inclusive communities, where "the participation and contribution of people living with dementia is encouraged, supported, and valued, and the care and support provided within the community is culturally safe, culturally appropriate, and mindful of diversity" (Public Health Agency of Canada, 2019, p. 31). Like the age-friendly communities movement, dementia-friendly initiatives shift the discourse around dementia from a deficit narrative of loss, decline, and care to a more holistic emphasis on well-being, participation, and social inclusion (Hebert and Scales, 2019).

As the population ages, social inclusion has become a more popular focus of public policy. However, the development of lasting institutions, policies, and programs to support social inclusion has been slowed by the perceived threat of population ageing and the potential costs of such changes (Phillipson, 2020). The *Improving Social Inclusion for People with Dementia and Carers through Sharing Dance* project sought to explore the potential of an arts-based program to support social inclusion. In doing so, we aimed to contribute to the evidence base that supports policies and practices as well as extend theoretical understandings of social inclusion. Many of the chapters of this book focus on different dimensions of social inclusion. In this chapter, we briefly highlight how the arts-based program shed light on and deepened our understanding of these different dimensions of social inclusion through a series of examples.

Conceptualizing age and dementia-related social inclusion

There is no clear, widely accepted definition of social inclusion (Pinkert et al., 2019). The lack of consensus in defining and conceptualizing social inclusion can be attributed to differences in disciplinary perspectives, the challenges of designing studies that capture such a multifaceted social phenomenon, and the difficulties of analysing as well as integrating complex sources of data to better understand social inclusion. Although different studies may focus on specific dimensions of social inclusion in later life, there are a series of common assumptions across studies of social inclusion that have guided our research on Sharing Dance Older Adults.

At the broadest level, the notion of social inclusion is construed as a fundamental value to guide policy and programs. This is reflected in the international and national policies cited above in which creating inclusive societies is described as a human right. Specifically, social inclusion highlights older adults' right to participate free from discrimination and the need to address ageism and stigma, including in relation to dementia (Government of Canada, 2017). For example, the stigma associated with dementia can prevent people from participating in opportunities, jeopardizing the right to participate in society. In addition, for organizations, communities, institutions, and governments to support opportunities for older adults, especially people living with dementia, they need to recognize, respect, and value their abilities and contributions (Herron and Rosenberg, 2017). Although respect and recognition of the rights of older adults is a foundation of the concept of social inclusion, most research focuses on individual experiences of inclusion with less attention to rights at a broader level.

Individual experiences of meaningful participation

Social inclusion is most often studied at the individual level. Most studies emphasize that social inclusion involves meaningful participation in social life. Meaningful participation is highly individual; as such, it is essential to centre individual experiences, values, and meanings to understand inclusion. By defining inclusion in this way, studies recognize the diversity of older adults in terms of their individual characteristics, interests, and social location; older adults have access to different resources to navigate inclusion and older adults may ascribe different values and meanings to various relationships and places. The focus on meaningful experiences sets inclusion apart from concepts such as integration. It suggests that interacting with others and participating in activities are part of the process of social inclusion, but they are not the final outcome. Inclusion can involve experiences of social connection (Moody and Phinney, 2012), social support (Dow et al., 2011), comfort, and belonging (Bartlett, 2007) in interactions with others. There is also an extensive body of research on social exclusion and what it means to be socially isolated or lonely (Cloutier-Fisher et al., 2011; Finlay and Kobayashi, 2018; Grenier and Guberman, 2009; Scharf et al., 2001).

The role of space and place

To address individual experiences of social inclusion or exclusion, research has emphasized the importance of modifying physical and social environments (e.g., age-friendly cities movement, dementia-friendly initiatives), recognizing that social inclusion is shaped by features of space and place. That is to say that the way an individual's environment is organized (i.e., space) and the feelings and meanings they associate with that environment (i.e., place) influence their experiences of inclusion and exclusion. For example, studies have examined how the organization of a room, household, community, or facility can influence the accessibility of services as well as relationships with others (Herron et al., 2020; Wiles, 2003). In addition, the social norms and routines within places shape inclusion and exclusion. For example, various studies have examined how familiarity with a community, lifelong routines, and strong social bonds can influence social inclusion in complex ways as one ages (Clarke and Bailey, 2016; Forbes et al., 2011). Sometimes these relationships within place can be very supportive and enhance one's sense of belonging; however, they can also contribute to feelings of exclusion when an individual is no longer able to participate in the same ways they used to in the past. This example illustrates the complex and dynamic nature of social inclusion as a process.

Process and outcome

Social inclusion is a process and an outcome. That is to say, once achieved or experienced, social inclusion is not a constant state of being. Experiences of inclusion and exclusion are constantly changing as individuals negotiate different relationships within their environment. Thus, research needs to explore the dynamics of inclusion and exclusion as well as moments where inclusion is observed. Studies situated within an interventionist paradigm, often focus on measuring or observing the number of social contacts people develop within programs as well as how programs influence individual perceptions of loneliness and caregiver burden (Camic et al., 2014; Dow et al., 2011). In this context, programs are viewed as instruments to change feelings and experiences as well as relationships. There is a lack of in-depth exploration of how programs actually support social inclusion over time and there is limited consideration of broader processes that enable inclusion (e.g., opportunities to provide meaningful feedback about a program or how a program might address accessibility challenges).

Multi-dimensional frameworks

Recognizing the dynamic nature of social inclusion, researchers have recently attempted to develop multi-dimensional conceptual frameworks of social exclusion and inclusion that describe the interaction of micro-, meso-, and macro-level influences on the social lives of older adults (Pinkert et al., 2019;

Walsh et al., 2019, 2021). For example, in a review of theoretical frameworks for dementia-related social inclusion, Pinkert et al. (2019) identified four dimensions of social inclusion common in the literature: emotional dimensions (e.g., feelings of comfort and belonging), environmental dimensions (e.g., community dynamics and institutional practices), economic dimensions (e.g., funding, and staffing healthcare systems), and cultural/societal dimensions (e.g., stigma). In general, literature on dementia-related social inclusion focuses on program and service environments, long-term residential care (LTRC), and community settings. For example, research on residential care environments has emphasized the ways in which these environments distance older adults from important relationships and often fail to provide enough private space, communal space, and means of transportation to support a range of meaningful interactions with others (Bartlett, 2007). Other research has identified the importance of maintaining quality relationships with friends, family, and healthcare professionals in promoting social inclusion and maintaining one's sense of identity and belonging (Forbes et al., 2011). Research on social and cultural dimensions of inclusion has tended to focus on stigma within community settings as well as health and LTRC environments (Bartlett, 2007; Forbes et al., 2011) and the ways in which it deprives people living with dementia of their human rights by restricting their involvement in their own care and well-being (Kontos et al., 2020). Less attention has been paid to structures and processes of inclusion beyond the immediate environments where people with dementia live or access care.

Notably, Walsh et al.'s (2019, 2021) framework for age-related social exclusion is more comprehensive in terms of the range of influences considered when compared to much of the social inclusion literature, particularly dementia-related social inclusion (Pinker et al., 2019). It includes six domains: social relations; services, amenities, and mobility; material and financial resources; neighbourhood and community characteristics; civic participation; and social and cultural influences. The individual and their experiences are at the centre of this framework, but they are shaped by an extensive range of structures and systems that surround them both within and beyond their immediate environment, including influences at the global, state, regional, community, and household level. For instance, the particular dimensions and their influence on social exclusion and inclusion may differ between countries because of the variation in policies, institutions, organizations, and societal values. Moreover, Walsh and colleagues' framework (2019) provides a multi-systemic approach to understanding and addressing exclusion that is useful when considering the broad range of factors that might promote social inclusion. Drawing together these multi-systemic frameworks, we defined social inclusion as an outcome and a process that involves supporting the rights and meaningful participation of older adults across a range of settings and scales.

Drawing these assumptions and limitations together, our research on the Sharing Dance program centred on individual experiences and meanings through the use of multiple qualitative methods (see Skinner and Bar, Chapter 3) that captured experiences at the beginning, throughout, and at the end of the program. This approach enabled us to be attentive to how rights and respect were recognized through the program design, in interactions with others, and from the perspectives of different stakeholders involved in the program. Recognizing the multi-dimensional nature of the social inclusion, we sought to understand the experiences and perspectives of different actors shaping inclusion processes including organizations hosting the program, family carers, and older adults. Our collaborative approach was critical to observing and documenting these diverse perspectives. In addition, we explored the role of multiple environments including various LTRC facilities and community settings in different communities and provincial jurisdictions. In doing so, we aimed to contribute to multi-dimensional frameworks for social inclusion in relation to ageing and dementia with an attentiveness to different experiences, positions, and environments.

In the section that follows, drawing on findings from the *Improving Social Inclusion for People with Dementia and Carers through Sharing Dance* project (see Skinner and Bar, Chapter 3), we explore individual, relational, programmatic, community, and institutional dimensions of social inclusion. We begin by looking at the importance of creativity and self-exploration in supporting emotional dimensions of social inclusion and meaningful participation at an individual level. We link creativity and individual expression to interpersonal processes (i.e., enhanced communication and understanding) as well as programmatic features (i.e., routine, and non-judgemental facilitation). We discuss the role of local and national organizations in expanding opportunities and resources to promote inclusion. In doing so, we highlight the integration of different dimensions of social inclusion.

Supporting emotional dimensions of social inclusion through creativity and exploration

The Sharing Dance program illustrated the value of supporting creativity and self-exploration to enhance meaningful participation at an individual level as well as support the emotional dimensions of social inclusion. Multiple older adults participating in the program commented that Sharing Dance encouraged creativity as well as individual exploration. For example, one older adult commented that “It was encouraged to try your own thing or try something new or go a little bit farther.” Other participants added that opportunities to creatively diverge from the instruction made them feel good about themselves and helped them to overcome the difficulties of trying something new. One older adult explained,

> When we were doing that bicycle song, I lifted my seat up and did the bicycle thing with my feet off the ground – I felt very good about it. I’ve never been an athletic kind of person and learning something new was kind of difficult for me. I just sort of complemented myself on my creativity – felt good about that.

The participant noted that she did not have to be an "athletic kind of person" to participate, suggesting that her sense of identity was not challenged by the program. She felt good and encouraged to participate in her own way. Similarly, another older adult living in LTRC, explained, "You could be yourself" and he said the program offered "Encouragement to do the things that [he]… had been doing" in reference to a narrative dance about fishing. In this way, the program affirmed participants' sense of identity and individuality.

Facilitators in the class learned from observing each individual's interpretation and expression and some were surprised to see so much individuality among participants. For example, one administrator in LTRC explained:

> I was surprised that everybody – everybody was kind of dancing their own interpretation whether we were fishing or riding their bike the way they remembered fishing, or riding their bike… everybody had their own little thing that they brought to it – the fishing one and rowing the boat – oh there's a beaver over there or swatting mosquitos – everybody had their own way of swatting a mosquito – their own expressions of their life and skills.

Examples of individual creativity highlight the importance of narrative dances for facilitating individual expression and identity as well as supporting relationality and meaningful interaction with others. As discussed further by Pia Kontos and Alisa Grigorovich in Chapter 7, the creativity that the program encouraged also supported and enriched relationships by fostering embodied expression. Indeed, this highlights the connections between individual and interpersonal processes of inclusion. That is to say that fostering individual comfort and confidence to express oneself and try new things influenced how one was understood and could relate to others.

Enhancing communication and understanding in relationships

Through repeated interactions in the program, family, staff, and volunteers gained a new awareness of the meaning of participants' gestures, which helped to build a greater understanding of others and their expression of different feelings. For example, during the first week of observation, a participant living with dementia in a LTRC made a movement during the program that was initially interpreted as a sign of anger or frustration. He shook his head and seemed to raise his fist. He repeated this move throughout his participation in the program in different contexts, which helped others in the room to deepen their understanding of this movement. The following researcher field note captures a moment in context and appears to reveal the participant's intent:

> Moving his feet around in time with the music, his hands shaking in front, the participant looks around to others in the room, and as another participant moves her arms like the instructor beside him, he motions with his fist in her direction… As applause sounds and the sequence ends, he clasps his hands in front and looks around at others smiling. The

> remote instructor says let's do it again, but faster, He picks up the "out and in" steps swiftly. On the remote instructor's direction "side touch, side touch," he looks over to his left at another participant, and moves his feet, it looks like he's wiggling his feet and he raises his fist at the screen.

The observation above outlines the interactions, facial expressions, and context in which this move occurred, suggesting that it was a sign of enthusiasm. During a follow-up interview with a staff member from the LTRC facility, she indicated that the program had helped her to understand the participant in a new way:

> The gestures [they] were making during the dance program suggested something different than [what] we were interpreting before – something that we thought was showing anger and frustration was actually showing excitement and engagement… as staff we could now support and let an outsider know – 'that's ok he's just saying hello'.

Importantly, the dance program deepened staff understanding of the participant in ways that challenged initial assumptions and allowed staff to better facilitate his interactions and relationships with others.

As another example, one family carer mentioned that she was surprised at what the person she cared for could do: "For me, seeing what you could do when you weren't able to stand and walk and dance – you can still do the arm movements and feet tapping. It's really refreshing to know that people still enjoy doing that…" Similarly, a volunteer at an LTRC residence reflected on her experience in the following way:

> Nice to see people interacting that don't interact. I would sit with Ralph quite a bit and you know I was very impressed with him that he followed and was taking notice and even smiling and enjoying it. And others would smile – it would bring something positive. You see a different side of people. They seemed happy and engaged.

One community participant said (of a fellow participant living with dementia) "Their expressions, camaraderie, how they participated opened my eyes that dementia shouldn't be a word not said." While these comments reveal how some participants, carers, and volunteers held stigmatizing views of dementia, they also suggest that the interaction, expression, and enjoyment associated with dance can challenge those views and foster greater understanding as well as respect for their abilities.

Change over time through routine and non-judgemental facilitation

Older adults, volunteers, carers, and staff were all very aware of the importance of time and its influence on the success of the program. They spoke to the importance of routine and repetition in building comfort and familiarity over time and they identified changes over time such as increased participation, increased relaxation, and increased communication. Participants noted that following a similar dance routine or protocol helped to facilitate individual growth and comfort, which in turn influenced their openness to others. They explained that they felt more relaxed, and the volunteers and staff observed that they looked more relaxed and comfortable. In one focus group in LTRC, an older adult said, "Well the repetition helps …you've done that already so you know how to do that part and then you go onto another part and you've done that. So … that's good about the repetition." Other older adults in a different LTRC facility agreed that they were "kind of self-conscious" at the beginning but this changed over time through routine, repetition, and particular features of the program.

Some participants attributed increased interaction over time to the facilitation and the environment it fostered. One older adult who participated at one of the active living centres explained, "I found that halfway through people were interacting more… Just the no pressure, you were allowed to have fun." She went on to add that as the program went on, she felt "Far more relaxed – that you could do anything you wanted as it progressed. And nobody was pointing and making fun of you if you screw up. You could chuckle and carry on." She emphasized the non-judgemental environment as an important feature. Like others, she pointed out that this environment was created over time through repeated interaction and emphasis on fun and individual expression and exploration.

A community-dwelling participant living with dementia summarized her experience of the program saying,

> Well we all knew each other by the end. We had a little visit as we went along. Being with a group is better than being by yourself. I enjoyed it and the people there were easy to get along with… near the end we were all getting better at it and that made quite a difference.

She valued the opportunity to get to know people through the program, and she felt that she and others experienced growth over time in terms of what they were able to do together.

Facilitators corroborated these experiences; they observed more engagement, comfort, conversation, and independence as the program progressed. One community facilitator stated,

> I think the engagement varied in at the start it was very hesitant, but by the third or fourth session that changed more as people became more comfortable with the program. They didn't need me as much, they were comfortable to participate on their own. Other programs people count on me more to lead.

Interestingly, the facilitator noted that as time progressed, participants directed themselves and they became more comfortable with one another.

Moreover, the supportive environment the program facilitated was produced over time, highlighting the importance of regular and sustainable arts-based programming to promote processes of social inclusion.

Expanding opportunities for social engagement in the community

Across the different sites, the program helped to provide opportunities for social engagement with a wider community. Each site approached program delivery differently with some LTRC facilities inviting community members, carers, and volunteers into the program and other community organizations inviting LTRC residents to attend the program as an outing in the community. In general, participants in both community and LTRC settings found getting out of their rooms or homes to be a very valuable part of the experience that increased the number of people they saw and provided them with opportunities to enact social ties through activity with others. For example, a participant living in a LTRC explained, "For myself I got to meet more people from the community, which is good." Another community-dwelling participant living with dementia said, "Well it was nice to get out and know that that day I had to get out and get somewhere." Each of these comments reflects place-specific challenges and opportunities to promoting social engagement. For the participant who lived in a LTRC – a place often associated with separation from meaningful relationships and community (Bartlett, 2007; Theurer et al., 2015) – the program temporarily challenged these meanings by bringing more volunteers, family members, and other community-dwelling participants into her LTRC facility and expanding opportunities for social engagement with others from outside the facility. For the community-dwelling participant, the program addressed the lack of opportunities for people living with dementia in the community and, similarly, provided a routine space for regular social engagement with others outside her house.

The opportunity to get out and meet new people was not exclusive to older adults living in LTRC nor those living with dementia. One volunteer explained,

> I'm not from this area, so I don't have relatives – I know [another participant] and a few from around town. It [the personal care home] was a place I would never go, so I was glad to have that chance to be part of it.

The volunteer's experience illustrates how different groups of older adults with various abilities and interests were drawn together through the program expanding social networks and resources among older adults in the community. As another example of expanding community networks and resources, one facilitator noted "I think by the third event [they were] more open to chatting and talking and learning about [our organization]." From his perspective, offering a new program helped to engage new members with other community resources the facility had to offer.

For some individuals, the sense of belonging to the community was fostered through a recognition of community culture. Information about music preferences as well as activities and events happening in the community or LTRC were collected by an on-site facilitator who provided it to remote instructors through an online portal. Both the process of soliciting group feedback and effectively integrating it into future sessions supported a sense of community belonging. For example, one volunteer in LTRC noted that after the dance session, "People would talk about the dance they used to go to – square dancing, their partner, growing up… It made people communicate because they had shared experience." In another community site, the participants discussed their disappointment that their local song request was not honoured by the remote instructor, but this prompted staff and participants to sing together the lyrics to the song that was requested and give each other recognition for their request. These examples highlight the importance of learning about and integrating community preferences to promote a sense of belonging as well as understanding on multiple levels – in the community, in the program space, and to the remote instructor.

"Part of something bigger"

Although macro-level processes of social inclusion are less evident in some of our data, there are examples where the program influenced larger-scale systems and processes in LTRC and rural communities. These examples highlight the potential of remote programming to augment community and institutional resources. For example, in LTRC, staff commented that the remote delivery of the program (which they were not responsible for) enabled staff to focus on developing relationships. There was a general consensus among staff that they, "…got to know residents better" because the remote delivery of the program provided "…a time to be interactive with residents in a new role." Staff emphasized that having time and the ability to focus on the residents rather than programming shifted the generally task-oriented nature of their work towards the people they were working with. This example highlights the important role of leveraging external resources – in this case, arts-based programming from outside the institution – to support more relational care within LTRC settings.

Leveraging external resources also emerged as an important theme at the community level. One community participant described the benefits of the Sharing Dance program saying, "It's nice to be part of something bigger." Others in the group agreed that being a part of a program spanning different regions made them feel that they, and their community, were important. Some older adults explained that having opportunities to participate in a national program was particularly important for older adults living in rural communities because "Small towns are having a real struggle to keep … to keep going." The same participant described being able to participate in Sharing Dance as an opportunity to "Sort of put us back on the map." Comments like these indicate that having access to national programming has particular meaning for older adults and rural communities. Although not without limitations, remote programming can partially address specific challenges related to rural exclusion. The technological limitations and rural development issues will be touched on in greater depth by An Kosurko, Ilkka Arminen, and Melisa Stevanovic in Chapter 8 and Verena Menec, Mark Skinner, and An Kosurko in Chapter 9.

Discussion and concluding comments

In this chapter, we aimed to illustrate how arts-based programs can deepen our understanding of processes and dimensions of social inclusion. Drawing on multi-dimensional conceptualizations of social inclusion, we began our exploration of Sharing Dance with a focus on what individuals gained or valued about the program and how that supported their meaningful participation. We found that encouraging individual creativity, exploration, and expression was critical to promoting meaningful and safe participation at the individual level. When individuals felt comfortable and confident to express themselves, this also influenced how other participants understood them. Overall, Sharing Dance illustrates the particular strength of the arts in supporting individual expression as well as enriching understanding and communication with others over time, which is critical to addressing experiences of social exclusion (Forbes et al., 2011; Kontos et al., 2020).

Meaningful participation did not just take place, it also took time. Older adults, volunteers, and staff explained the importance of routine and repetition in building individual comfort and confidence as well as enabling social engagement. At a programmatic level, the dance protocol as well as skilled, non-judgemental facilitators helped promote social inclusion over time. Conceptually, time is an important and sometimes neglected feature of processes of social inclusion that deserves more attention in future research. Practically, supporting routine and lasting arts programming in community and institutional settings is critical to promoting social inclusion in the long term.

The Sharing Dance program expanded the range of places in which various older adults enacted social ties, expanding community social networks and a sense of belonging. The increasing number of places and people participants came to know was not just an important outcome; it revealed a number of processes critical to supporting these ties. Collaboration between community organizations was essential to bringing people together to participate through community program promotion as well as transportation. Although not without challenges, organizations were able to bridge the distance often created by relocation to residential care by transporting residents outside the facility or inviting community members to join in the facility. In addition, collaboration and information sharing with Canada's National Ballet School (NBS) was critical to ensuring the program was culturally appropriate and responsive to community practices and preferences.

The examples in this chapter illustrate how processes and resources across multiple dimensions can come together to support social inclusion of older adults. For diverse older adults to experience respect and recognition of their abilities, they need opportunities that support their individual creativity and exploration. Engaging in these opportunities with others in the community can further deepen understandings and enhance respect for the diverse abilities of older adults. When others see older adults grow and express themselves, this can challenge stigmatizing views within the context of the program environment. For such an interaction to take place, community organizations and institutions as well as national policies and strategies must support sustainable arts-based programming. Yet, the rights of older adults to accessible and sustainable arts-based programming are often neglected or viewed as secondary to other health, social, and community resources. Importantly, as both theory and the examples from Sharing Dance suggest, resources to support social inclusion can be mobilized across multiple dimensions. By fostering creativity, understanding, non-judgemental environments, and regular opportunities for social engagement, community belonging, and collaboration across multiple jurisdictions arts-based programs can help promote social inclusion in practice as well as advance multi-dimensional models.

References

Bartlett, R. (2007). 'You can get in alright but you can't get out': Social exclusion and men with dementia in nursing homes: Insights from a single case study. *Quality in Ageing and Older Adults*, *8*(2), 16–26. https://doi.org/10.1108/14717794200700009

Bartlett, R., and O'Connor, D. (2010). *Broadening the Dementia Debate: Towards Social Citizenship*. London: Policy Press.

Camic, P. M., Tischler, V., and Pearman, C. H. (2014). Viewing and making art together: A multi-session art-gallery-based intervention for people with dementia and their carers. *Aging & Mental Health*, *18*(2), 161–168. https://doi.org/10.1080/13607863.2013.818101

Clarke, C., and Bailey, C. (2016). Narrative citizenship resilience and inclusion with dementia: On the inside or on the outside of physical and social places. *Dementia*, *15*(3), 434–452. https://doi.org/10.1177/1471301216639736

Cloutier-Fisher, D., Kobayashi, K., and Smith, A. (2011). The subjective dimension of social isolation: A qualitative investigation of older adults' experiences in small social support networks. *Journal of Aging Studies*, *25*(4), 407–414. https://doi.org/10.1016/j.jaging.2011.03.012

Courtin, E., and Knapp, M. (2015). Social isolation, loneliness and health in old age: A scoping review. *Health & Social Care in the Community*, *25*(3), 799–812. https://doi.org/10.1111/hsc.12311

Dow, B., Haralambous, B., Hempton, C., Hunt, S., and Calleja, D. (2011). Evaluation of Alzheimer's Australia Vic Memory Lane Cafés. *International Psychogeriatrics*, *23*(2), 246–255. https://doi.org/10.1017/s1041610210001560

Finlay, J. M., and Kobayashi, L. C. (2018). Social isolation and loneliness in later life: A parallel convergent mixed-methods case study of older adults and their residential contexts in the Minneapolis metropolitan area, USA. *Social Science & Medicine*, 208, 25–33. https://doi.org/10.1016/j.socscimed.2018.05.010

Forbes, D., Ward-Griffin, C., Kloseck, M., Mendelsohn, M., St-Amant, O., DeForge, R., and Clark, K. (2011). 'Her world gets smaller and smaller with nothing to look forward to': Dimensions of social inclusion and exclusion among rural dementia care networks. *Online Journal of Rural Nursing & Health Care*, *11*(2), 27–42. https://doi.org/10.14574/ojrnhc.v11i2.18

Government of Canada. (2017). Social inclusion of individuals with dementia and caregivers. https://www.canada.ca/en/institutes-health-research/news/2017/02/social-inclusion-individuals-dementia-caregivers.html

Grenier, A. M., and Guberman, N. (2009). Creating and sustaining disadvantage: The relevance of a social exclusion framework. *Health & Social Care in the Community, 17*(2), 116–124. https://doi.org/10.1111/j.1365-2524.2007.00804.x

Hebert, C. A., and Scales, K. (2019). Dementia friendly initiatives: A state of the science review. *Dementia, 18*(5), 1858–1895. https://doi.org/10.1177/1471301217731433

Herron, R. V., and Rosenberg, M. W. (2017). "Not there yet:" Examining community support from the perspective of people with dementia and their partners in care. *Social Science & Medicine*, 173, 81–87. https://doi.org/10.1016/j.socscimed.2016.11.041

Herron, R. V., Funk, L. M., Spencer, D., and Wrathall, M. (2020). Assisted living facilities as sites of encounter: Implications for older adults' experiences of inclusion and exclusion. *Ageing & Society*, *40*(7), 1577–1593. https://doi.org/10.1017/S0144686X19000187

Holt-Lunstad, J., Smith, T. B., Baker, M., Harris, T., and Stephenson, D. (2015). Loneliness and social isolation as risk factors for mortality: A meta-analytic review. *Perspectives on Psychological Science*, *10*(2), 227–237. https://doi.org/10.1177/1745691614568352

Hung, L., Leitch, S., Hung, R., and Phinney, A. (2020). Creating dementia-friendly and inclusive communities for social inclusion: A scoping review protocol. *BMJ Open*, *10*(6), e035028–e035028. https://doi.org/10.1136/bmjopen-2019-035028

Kontos, P., Grigorovich, A., Dupuis, S., Jonas-Simpson, C., Mitchell G., and Gray, J. (2020) Raising the curtain on stigma associated with dementia: Fostering a new cultural imaginary for a more inclusive society. *Critical Public Health*, *30*(1), 91–102. http://doi.org/10.1080/09581596.2018.1508822

Leigh-Hunt, N., Bagguley, D., Bash, K., Turner, V., Turnbull, S., Valtorta, N., and Caan, W. (2017). An overview of systematic reviews on the public health consequences of social isolation and loneliness. *Public Health* (London), *152*, 157–171. https://doi.org/10.1016/j.puhe.2017.07.035

Moody, E., and Phinney, A. (2012). A community-engaged art program for older people: Fostering social inclusion. *Canadian Journal on Aging*, *31*(1), 55–64. https://doi.org/10.1017/S0714980811000596

Phillipson, C. (2020). Austerity and precarity: Individual and collective agency in later life. In A. Grenier, C. Phillipson, and R. Settersten (Eds.), *Precarity and Ageing: Understanding Insecurity and Risk in Later Life (Ageing in a Global Context)* (pp. 215–235). Chicago: Policy Press.

Pinkert, C., Köhler, K., Von Kutzleben, M., Hochgräber, I., Cavazzini, C., Völz, S., … Holle, B. (2019). Social inclusion of people with dementia – An integrative review of theoretical frameworks, methods and findings in empirical studies. *Ageing & Society*, *41*(4), 773–793. https://doi.org/10.1017/s0144686x19001338

Public Health Agency of Canada. (2019). *A Dementia Strategy for Canada*. https://www.canada.ca/content/dam/phac-aspc/images/services/publications/diseases-conditions/dementia-strategy/National%20Dementia%20Strategy_ENG.pdf

Scharf, T., Phillipson, C., Kingston, P., and Smith, A. E. (2001). Social exclusion and older people: Exploring the connections. *Education and Ageing*, *16*(3), 303–320.

Shakespeare, T., Zeilig, H., and Mittler, P. (2019). Rights in mind: Thinking differently about dementia and disability. *Dementia*, *18*(3), 1075–1088. https://doi.org/10.1177/1471301217701506

Swaffer, K. (2015). Dementia and Prescribed Disengagement™. *Dementia, 14*(1), 3–6. https://doi.org/10.1177/1471301214548136

Theurer, K., Mortenson, W. B., Stone, R., Suto, M., Timonen, V., and Rozanova, J. (2015). The need for a social revolution in residential care. *Journal of Aging Studies*, 35, 201–210. http://dx.doi.org/10.1016/j.jaging.2015.08.011

Walsh, K., O'Shea, E., and Scharf, T. (2019). Rural old-age social exclusion: A conceptual framework on mediators of exclusion across the lifecourse. *Ageing & Society*, *40*(11), 2311–2337. https://doi.org/10.1017/S0144686X19000606

Walsh, K., Scharf, T., Van Regenmortel, S., and Wanka, A. (2021). *Social Exclusion in Later Life: Interdisciplinary and Policy Perspectives*. Cham: Springer International Publishing AG.

Wiersma, E. C., and Denton, A. (2016). From social network to safety net: Dementia-friendly communities in rural northern Ontario. *Dementia* (London, England), *15*(1), 51–68. https://doi.org/10.1177/1471301213516118

Wiles, J. (2003). Daily geographies of caregivers: mobility, routine, scale. *Social Science & Medicine*, *57*(7), 1307–1325.

World Health Organization (WHO). (2007). *Global age-friendly cities: A guide*. Geneva: World Health Organization. https://apps.who.int/iris/handle/10665/43755

World Health Organization (WHO). (2012). *Dementia: A public health priority*. World Health Organization. https://www.who.int/publications/i/item/dementia-a-public-health-priority

World Health Organization (WHO). (2017). *Global action plan on the public health response to dementia 2017–2025*. World Health Organization. https://www.who.int/publications/i/item/global-action-plan-on-the-public-health-response-to-dementia-2017---2025

World Health Organization (WHO). (2021). *Global report on ageism*. Geneva: World Health Organization. https://www.who.int/publications/i/item/9789240016866

6 Voices and lived experiences of *Sharing Dance*

Rachel Bar, Margaret Dunphy, Rachel Herron, Ruth Snider, Whitney Strachan, and Craig Wingrove

Introduction

In research on ageing and dementia, the inclusion of lived experience perspectives has become essential to deepening understandings of the daily life, choices, and values of diverse older people (Sabat, 2001). Rather than view ageing and dementia as problems to be solved, research on the lived experiences of older adults illuminates the capabilities, agency, and interests of older adults as they navigate the everyday world (Harris, 2008; Herron and Rosenberg, 2017; Wiles et al., 2012). Responding to advocacy movements calling for "nothing about us without us," (Bryden, 2016, p. 1) scholars of ageing acknowledge that the voices of older adults are critical to developing programs and policies that enrich their lives (Novek and Wilkinson, 2019). In addition, the voices and perspectives of those that support and deliver programs are also critical to understanding the dynamic and relational experiences that enable programs to succeed in fulfilling the desires and interests of older adults.

The aim of this chapter is to centre the first-hand experiences of different stakeholders within a collaborative arts-based research project. In the case of the Sharing Dance Older Adults program and the research that shaped the program, there are many different voices and experiences to consider. Each individual's experience is shaped by different factors including their role in the program as a participant, artist, or facilitator as well as their social identity (e.g., age, gender, ability, and race) and past experiences outside the program. Conceptually, different viewpoints were critical to exploring the impact of Sharing Dance in relation to ageing, social inclusion, and care. Each of these concepts involves considering how experiences converge or diverge in ways that are supportive or contradictory. Methodologically, it has been critical to consider multiple and varied voices throughout the research project and its many phases (see Skinner and Bar, Chapter 3). As the research team disseminated research findings, we wanted to continue to centre the voices and experiences of different collaborators and participants. In doing so, we sought to provide a more comprehensive, empathetic, and rich understanding of the program and its impact on those involved.

DOI: 10.4324/9781003260691-6

This chapter includes the perspective of pianist Craig Wingrove (from Canada's National Ballet School (NBS) in Toronto, Ontario), Sharing Dance participants Ruth Snider (from Ennismore, Ontario) and Margaret (Margo) Dunphy (from Minnedosa, Manitoba), and community facilitator Whitney Strachan (from the Alzheimer Society of Manitoba in Brandon). To prepare this chapter, Rachel Bar and Rachel Herron reached out to each of the contributors to ask them if they would contribute to writing a book chapter about experiences of Sharing Dance. To facilitate the process of writing the chapter, we met with each of the authors virtually and asked them a series of similar questions. We asked them to tell us about themselves and their involvement in Sharing Dance; to describe their experience of the program over time; to describe what Sharing Dance meant to them; and to consider what other artists, facilitators, and participants should know about the program. With their permission, we recorded these conversations and used the transcripts to develop a narrative, which we shared with each participant to provide questions, edits, critique, and affirmation. Although each of the chapter contributors had different roles and experiences in the program, their narratives revealed considerable convergence around shared experiences of joy, social connection, and growth and learning. In the sections that follow, each contributor shares their voice and experience within the program, while Rachel Bar and Rachel Herron reflect on the themes that emerged from each story.

Craig's voice

I am musician. I started playing the piano for dance classes at NBS, in Toronto, in 1991. Over the past 30 years, I have played for the professional ballet program, and for the School's child and adult recreational programs. I've also been teaching music for the last 15 years in NBS' dance teacher training program, and then I got involved in Sharing Dance. Playing as the accompanying pianist for Sharing Dance classes has been a lot of fun and I just love everyone who's involved in it. There's a lot of positive energy going on.

Part of playing for these classes is playing songs that I feel people would like from different genres, there's pop, rock, jazz – usually fun stuff. But also sometimes the teachers will make a story or they'll have a theme, like a beach. I have to take the dancers on a journey with my improvising. I have to start telling a story that matches what the teacher's doing and make it match through leg lifts and their upper body and try and get the feeling of the waves and the beach, or waves in the ocean. So improvising is a huge part of this because it's alive and you can take the people there on a journey.

My first Sharing Dance classes were recorded in a tiny practice studio with a wonderful teacher teaching the classes to a group in Peterborough. I didn't really understand what it was all about yet; it was all so new. After we recorded a bunch of classes we said, "Oh, let's go visit the dancers in Peterborough." Meeting participants in-person for the first time, I just remember how kind of famous it seemed we had become in those peoples' eyes. We walked into the room and heard – "Oh my god, she's here! He's here! They're here! It's CRAIG!" You don't expect people to grab onto you. It's not just the movement people latch onto. They latch onto the people. They were all getting together to do this program and I was blown away by how much it meant to them. It took on a power that I was not expecting at all. Sharing Dance classes gave them a reason to get involved and have some fun. It highlighted for me how important the arts are for supporting quality of life. I know this even just from the amount of joy that I've seen in these classes.

The words that come to mind when I'm playing for a class are fun, positivity, physicality-focused, and artistry. The artistry part, a lot of people are missing in their lives. It is so important to be able to express yourself through physical activity. It does something to your body and to your brain. I know this just from watching this go on. I thought I would just be playing, but as an artist, I was just amazed at the amount of artistry that was coming out because I was getting such feedback from them. I wasn't just playing the piano anymore. I realized the quality of the music that I played can make your body move forward or move sideways or it can make you do different things. I'm not just playing notes – I'm playing notes to get into your heart and into your muscles. It's really what it's all about, and that caught me by surprise. I was surprised by the level of artistry and how important it is. I saw people emote and get into it and I can see it does something to them, and I feed off that. I noticed that even through the screen, people can feel your energy and your positiveness. I think people feel like you're talking because I really am when I play. I demand myself that I get into your heart, even if you're not there.

What I think makes everything so powerful in this program is the marriage of these art forms. It's the marriage of the movement and the music. When you add two art forms together and both musician and dancer are really working together, it's incredibly powerful. I've experienced it a lot; each art form by itself is good too, but when you add two, and they're both simpatico, it's incredibly powerful and uplifting. I realized also that when you play something that a person asked for, wow, it made them feel really special. We feel really special when someone listens to us, so it was a great chance for me to make someone feel heard and special.

Love was also a component of my experience playing for Sharing Dance. I really love all the teachers I was working with. When I was doing this live and I got to know the people a little bit they came up and talked to me during the breaks and after the class. You start to love them too. I love the teachers, I love the participants; they love me. When there's love in the room, here we go, now we're ready to rock the world. I think it all boils down to that love. I think that's the bottom line, really. It makes me feel like I'm doing something special and I'm helping people. We feel good when we feel what we do is important. I feel important. I noticed how much this was needed in people's lives. It's probably the situation where I really noticed the power of music and dance most (Figure 6.1).

If I was talking to another musician about Sharing Dance, I would want them to know their music really matters. Try and emulate the feeling that's needed

Figure 6.1 Craig Wingrove playing for a Sharing Dance Older Adults class.

with the phrasing, the articulation, the dynamics, learn a ton of songs and a ton of genres and also make sure your improvising is really, really, really good! Most importantly though I would say, please put emotion and put your heart into it. If you're not, I think we're just wasting our time.

Craig's reflection on his experience with Sharing Dance, as an artist, highlights how he grew and learned from his involvement in the program, the role of the arts in supporting emotional expression, and the importance of relationality in producing each dance session. He was not sure what to expect at the outset, but he grew and learned from his interactions with the teachers and dancers. It surprised Craig how much artistry was involved in Sharing Dance. He found his artistry was driven by feedback and learning how important the arts were for the dancers in the program. He observed that music helped make participants move in a certain way and he saw them express themselves and emote when he played in person. Even when he played for remote classes, his understanding of people's relationship with music shaped how he played. Getting to know what music the dancers like was important to him and so was playing music that tells the story of a dance and allows dancers to imagine themselves in that story. He felt a responsibility to get into people's hearts through music.

Part of what made being involved in the program meaningful for Craig was the relationships he developed with teachers and dancers. He also felt a sense of purpose in his interactions with dancers. He saw his music was helping people and it was important in people's lives. Dancers, like Ruth, shared their experience of growth and learning as well as a sense of connection with others through the program.

Ruth's voice

I'm Ruth Snider and I live in a rural area on a little island with a causeway. I love living in the country. When I was a pre-teen, I was told after an accident I could never do sports. I became a senior who had two hip replacements and started doing physio, tai-chi, and restorative yoga with this absolutely amazing group. Thankfully, NBS saw that there's lots of seniors in my area and this was a dynamic group so they piloted Sharing Dance with us. I have been with NBS from the beginning of Sharing Dance. My experience started by meeting Mark Skinner and the NBS crew in the room where we did yoga and from there it has expanded all across the country.

Sharing Dance is about mind, body, and soul. Lots of other programs have bits and pieces of that, but they don't have all of it. They may even have some mind, body, and soul pieces, but they don't have that connecting across for the whole country. Sharing Dance facilitates community in a much larger sense. Plus, I've learned that NBS' community is huge! The fact that you've allowed us to be a part of your unbelievable dance community is incredible and it is an honour to be a part of the impact Sharing Dance has on lives across the country.

I remember when they came out and interviewed me and a friend and the national news did a beautiful piece on Ennismore. That was wonderful because it gave a national perspective to what our little community was doing with the help of a national crew. There were follow-ups and celebrations in which you met lots of people, different organizations, as well as focus groups. A couple of times I and a few others were invited to NBS for events and that was awesome. It made me feel a part of the bigger community.

The reason Sharing Dance impacts me so much (and the people I have connected with) is that you're connecting with that person and you're like-minded in that you're getting together. You socialize in a way that's possible through technology even over a distance because you always have encouragement, whether it's with a facilitator or whoever. We were encouraged to tell NBS what we were thinking, to get in touch with them, what music we would like, what we are doing in our community, what bike-a-thon are you doing that week and give a shout-out too. They brought canoeing into our dance and all those things that mean a lot to people in my region.

It's about community, it's about bringing us together, it's about all of us older adults in rural areas who need to get out and connect. I've met so many people in greater depth across the Ennismore area. Improved connections are especially important for older adults, for rural people. When I broke my arm this year, other than my family, none of whom live really close to me, it's been the group I did Sharing Dance and yoga with who were there, and saying, "How are you doing? What can I do for you?" Those relationships grew from this experience. We still feel connected and we don't feel as isolated.

To find out in your sixties that there's more things to learn about yourself and to be aware of those abilities and have faith in them and also to nourish them and want to move further within that realm has been a benefit. I don't remember

Figure 6.2 Ruth Snider dancing with her community.

Ruth Snider, front row, first from the right.

when I first said I'm a dancer, but I didn't believe it. And now I say to people, "I dance. I dance in a different way, but I'm a dancer." I can't imagine me ten years ago saying to anyone I was a dancer but because of all of you, I am! And I can say that proudly (Figure 6.2).

Even my joy of watching ballet has grown through Sharing Dance. I share a lot of things with my older sister, and she said, " I never really got contemporary ballet." and I said, "Yeah, I don't think I did before either, but I do now, and I love it." So even those kinds of stretching of growth into areas of things that you saw but had less understanding of, it's enriching, it's joyful, it's wonderful. People noticed a difference in me and so my friends said, "I think I might do that." And the next friend said, "I think I might do that." I've had so many serendipitous moments with friends and people that I've just met and people I'm getting to know, who say I think I need to try that and then they do and they love it.

Participating in Sharing Dance has brought into my life the music, the movement, the dance, the camaraderie, the community, the research, all of these things have been important pieces in my life, and they're all coming together in my retirement years, which to me is amazing. I tell others about the joy I've had because of the friendships, because of the increased movement, control, and balance, because of the connections I've made in my body, in my mind, and in my soul. I have the joy of belonging to a world that lightens and brightens your day. These are the things that I have been able to receive from being involved in Sharing Dance. It's the connectedness and the joy, bringing your health in all aspects to a better place instead of becoming one of those little old grannies. It is one of those examples of something that can make our lives so much better in the world we live in, despite pandemics, despite living isolated in a rural area. It's the connectedness to a world that lightens and brightens the soul.

Ruth emphasized the holistic benefits of Sharing Dance as something that set it apart from other programs. For her, a critical feature of the program was the opportunity to be a part of a bigger community, a national community dance program within her own community. Having a voice in the development of the program was also important and seeing one's feedback incorporated into dance sessions contributed to feelings of community, belonging, and shared interests. Similar to other research on social inclusion in later life, Ruth identified a deepening of social connections in her community as one of the benefits of participating in the program. Her reflections suggest that the program challenged internalized ageist assumptions about older adults being unable to grow or learn new things; she identified a growing sense of confidence and pride as a benefit of the opportunities she had to grow and learn in the program. Like others we spoke to in developing this chapter, she used the word joy to describe her physical, social, and emotional experience of the program.

Whitney's voice

My name is Whitney Strachan and I work at the Alzheimer Society of Manitoba. I got involved with the Sharing Dance program and research when Rachel Herron reached out to us and our clientele and started asking questions about a new dance program for people with dementia like, "Hey, would this work? Do you think this is something we could possibly try? What are your thoughts on this?" Initially the role of the facilitator was not very well defined so it felt a little bit like walking in blind. What we thought in the beginning definitely wasn't where we ended up in the end.

I was really sceptical for my clients at first. I didn't know how well they would be able to participate and how much information they would be able to retain. Especially at the start, I was in protective mode of the clients. We were going to be with all these researchers. I didn't want my clients to feel like they were on display. After the demonstration session went better than any of us expected it to, and we got a better understanding of what this program might actually look like, I thought "Okay, we could do this." Then we talked about doing the program virtually and I thought, "Oh, no, I don't know if this is going to work anymore" because for some people with dementia, the ability to watch something on the TV, they just are not able to process it. That's when I felt more doubt or scepticism again, but we did a lot of fine-tuning and tweaking through all the different sessions and I think the end result was definitely worth it.

Sharing Dance was a way for my clients with dementia to express themselves and a way for them to reconnect with the people in their communities, their caregivers, and their partners on a different level. It's that spirit of being together. It brought a spark and enjoyment back to life. For the duration of a class, they were not dealing with the disease or all the things surrounding the disease, they we're just being themselves, enjoying themselves, enjoying the time that they were with the people that were around them. There was no judgement, no restrictions. For that time they were being themselves and enjoying it. They were able to be in a setting where there were no expectations set on them in any way, they could just be themselves and enjoy being in that environment and being with other people for that moment in time.

Caregivers experienced the joy of seeing that person happy, that person being themselves, being expressive, being in a group, being accepted, they didn't have those worries anymore. I think that's why the program brought them a lot of joy. They could relax, be themselves, and just enjoy being in the environment with other caregivers and being out in the community as well. They didn't have to worry about "making a scene" or other people in the community judging them or worrying that something is going to go wrong. They could stop being a caregiver and be their individual self. If it's a daughter looking after their mother, they didn't have to have that caregiver relationship anymore when they walked into the program. They could just be mother-daughter and enjoy the program. For caregivers, that was one of the things they enjoyed the most, and they saw the person living with dementia enjoying themselves. I think that brought them joy.

I would see participants struggle with a dance and then see them excel later with a similar dance. That brought a spark of joy. You're so proud and elated for that person for what they were accomplishing. I think that was one of the highlights for me of the program; seeing those people I work with every day in an environment where they were excelling. You could see a change from week one to week eight. By week eight, they were excited to come to the program, they were excited about what they were doing, there wasn't any fear or hesitation, and they were just enjoying, and there was so much more interaction with other members.

For myself, it was different from a regular counselling session or activity session. I was interacting with my clients in a different way. It was another way for me to engage with people living with dementia and get to know them and interact with them on a whole other level. It was almost like adding another layer or another level to their personality. You saw some things that were probably always there but never really brought out. You saw a lot more of that fun side of them coming out. For some of them you saw a little bit more of their cheeky side, where they're laughing, they're making some funny jokes and things like that. These are things that you might not normally have brought out in a normal conversation and just doing an activity together. You saw things coming to the surface in Sharing Dance and it was just a nice added part of their personalities to get to know.

The benefit for me and the Alzheimer Society being involved was that it broadened our knowledge of what a person with dementia's abilities can be. It opened up those doors to us when talking and speaking with clients that there's a lot that they're still able to do. Participating in the program broadened our knowledge and added to the services we provide.

For people in the community, it was an eye-opening experience for them to see what a person living with dementia is able to do and a different way of communicating with them. They saw, if something didn't go quite right or we weren't following along exactly, it was fine. You can just carry on, accept it, and go with them. You know, it's not something we have to stop, we can just continue on that journey with that person living with dementia and still have a wonderful time with them.

I saw the program evolve and improve. I think that was one of the highlights. About halfway through I thought, "Okay, yeah, I'm fully onboard with this." The communication back and forth and NBS' willingness to adapt to what some of our clients and what people in the community were wanting and needing helped. Seeing the program evolve to where we got in the end was wonderful.

Sharing Dance was a wonderful way for people with dementia to be expressive. What we thought was impossible or we were sceptical about, we saw kind of flipped on its head a bit. We saw people excel and thrive in their environment.

Like other stakeholders, Whitney expressed some uncertainty and scepticism when she was initially introduced to the idea of hosting the Sharing Dance program for people living with dementia in the community. Once she experienced the program, she saw what people living with dementia could do in the program and she saw the joy the program brought to people living with dementia, carers, and herself as a facilitator. She identified the non-judgemental environment the program created as a critical condition for supporting joy, relaxation, and self-expression. From Whitney's perspective, seeing people living with dementia express themselves within a safe environment fostered learning and growth among everyone. She observed people living with dementia growing and learning new things from week to week. She also observed different dimensions of the person through the expressive opportunities that the dance program offered. Seeing more of these people fostered learning and growth among others. It challenged stigmatizing views of people living with dementia that suggest they cannot grow and learn new things. It also provided different opportunities to relate to the person creatively through dance.

Margo's voice

My name is Margo Dunphy and I participated in the Sharing Dance program at Minnedosa 50+ centre. I originally started going to it because my sister was going and I thought she would feel more comfortable if I was there. She has dementia and sometimes gets confused or upset. I find that if I'm there, she'll turn to me for help. Her husband signed her up to give her something different to do and get her out and about. At the time, I was also the president of the 50 + club and I felt I should at least know what was going on.

When everybody started, we didn't know what we were getting into, but the music and the teacher certainly just made things awesome. It just made you want to participate because it was fun. The woman doing the instructions from Toronto was just awesome and she didn't talk down to anyone. She included everyone in the program. She was a very personable leader and you didn't feel like you were taking it long-distance. She talked to you like you were there, and you almost felt like you were. Both her and the pianist, they were awesome! I enjoyed the way the exercises were incorporated with the music and the pianist was more than willing to change the music to what the participants wanted. He made it fun. The instructor made it fun too. They just made everything that much more senior-friendly and just friendly all around.

The program was set up so that it didn't matter if you had a disability. People really got into the music and they did the best they could. They didn't feel like they were on display for other people to stare at. They just kind of got into the groove and did what they could do, and if they couldn't do it, they laughed it off and kept on going. Over time, I found that people would laugh more and they had more fun and put more into it. I felt the participants really enjoyed seeing when the others gradually opened up and started to enjoy the program, and I know they certainly enjoyed the beautiful music.

It was a good way to get my sister out to mingle with others. My sister's a people-person, and she just enjoyed seeing other people and she really got into the music. It meant a lot! It was stimulating and motivating. It just mattered that everyone was willing to participate. The purpose of Sharing Dance is to get people together. She was never in a hurry to go home. She always wanted to have a chit-chat. It gave her husband a break too.

I really enjoyed the camaraderie of everyone that participated. I found that as time went on, the people got a little bit more relaxed with each other. My personal benefit was making sure my sister was out and about, and she was having fun doing it. For my sister, it was just nice to see people out and being able to talk to people that she knew. I found a lot of participants would have a little chit-chat with Dawn and it was nice to see them out and chitchatting with each other before the dance started. A lot of people there don't get a chance to do that because they live alone. Sharing Dance really helped them be able to do that.

I knew most of the people that were there and just having somebody walk in the door and somebody say, "Hi Mary," or "Hi Joe," that makes all the difference because, hey, somebody knows me, and I'm here, they know I'm here. That's what Sharing Dance did for a lot of people, they realized they weren't alone and other people still remembered them and still wanted to talk to them.

I found for myself, it helped me do more exercising and I found I was doing it at home. I'd put the music on, sit in a chair and do some of the dance moves at home that I could do and it had my sister certainly do more exercise. I'd go over and we'd turn on the music and I'd get the old time station and we'd sit and relax and enjoy the music and sometimes I'd catch her singing along with it and moving her feet. My sister and I have always been close, and I would say this Sharing Dance just brought us a little closer. It gave her someone in the family she could do something with. We used to play cards together. Now music is the next best thing. She's always been very musical. Sometimes these songs bring back memories for her so that's kind of nice.

I want others to know that Sharing Dance is really helpful to people living with dementia and, if it was available, they should see how much they enjoy it. It brought my sister out of herself a lot. She was at the stage where she realized that she was forgetting things and it made her very self-conscious but I think Sharing Dance helped her accept it and she could have fun with it.

Margo's insights speak to the ways in which the program was fun, meaningful, and accessible for both her sister living with dementia and also for herself as a carer. Notably it was something that she and her sister could do together. When other activities were no longer accessible to her sister, the program offered a new way to enjoy their time together and socialize with others as well.

Further, Margo's reflections describe ways in which the program inspired the infusion of music and dance outside of the class setting. Sharing Dance inspired Margo to find ways to enjoy moving to music on her own time as well. It was not just an activity reserved for the one time a week she took her sister to classes. It helped her see the opportunity to experience the joy of dance on her own time.

Discussion

Craig, Ruth, Whitney, and Margo all describe the ways in which Sharing Dance provided unique opportunities. Whether it was the holistic benefits of movement and music or a new opportunity to interact with people in their community or people they cared for, each emphasized the ways in which the context of the program fostered new opportunities. Most notably, the non-judgemental, accessible, and creative context provided opportunities for joy, social engagement, learning, and growth.

Joyful experiences were described by each stakeholder in this chapter. For Craig, playing music for the dance classes brought him pleasure and he felt loved and appreciated in his role. He described the joy he saw in others participating in the classes as well. Ruth and Margo both emphasized the joy they experienced from the dance movements as well as the relationships they developed and enriched through Sharing Dance. For Whitney, there was joy in seeing her clients excel and she observed joy too as she watched people living with dementia and their carers together. Consistent with previous research about older adults' experiences of dance, Sharing Dance brought deep pleasure to participant's everyday lives (e.g., Houston and McGill, 2013; Schneider and McCoy, 2018). The very physical nature of dance (e.g., moving, touching) itself provided an opportunity to embody emotion (e.g., smiling and laughing) and share those experiences of joy. Craig, Ruth, Whitney, and Margo add to the dimensions in which we understand how dance may act as a facilitator of joy, thus highlighting further shared benefits of dance opportunities.

Every stakeholder also commented on the important social benefits of Sharing Dance, including providing opportunities for social engagement, enhanced social connectedness, fostering social support, and developing a stronger and broader sense of community. Older adult participants emphasized the value of having opportunities to enact social ties in programs and activities within their communities. From their perspective, the program addressed dimensions of social exclusion, specifically lack of social opportunities and accessible programs (Walsh et al., 2019). Consistent with previous research on social inclusion in later life, the program enabled older adults to connect in community in new ways including developing a stronger and wider sense of community (Moody and Phinney, 2012). Ruth described a sense of belonging and shared interests with the dance community she had become part of through her participation in the program. She also found that the program fostered social support networks that supported her outside the program. In addition, Margo and Whitney's perspectives extend existing research on the social and emotional benefits of art-based programming for people living with dementia and their carers (Kontos et al., 2021). They both experienced more meaningful and close connections with people living with dementia through the program.

Opportunities for social engagement were valued not only by older adults in the program, but also by artists and facilitators who met new people or developed a new understanding of people they had already known. Specifically, Craig and Whitney highlighted the important role of the arts in supporting meaningful interactions with others, whether through getting to know what music people like or seeing the way people emote or express themselves through music and dance. The narratives of Whitney and Craig reveal the reciprocal benefits of the sharing dance for various stakeholders, including artists, program facilitators, and community members; this is in contrast to some health interventions, which tend to focus on the outcomes of arts-based programs for older adults with less consideration of the reciprocal benefits for those involved in delivering and supporting the program. Understanding the benefits for all is critical to promoting enriched relations of care as well as supporting social inclusion.

Finally, every stakeholder in this chapter spoke of learning something new about themselves and others through Sharing Dance, thus supporting opportunity for human flourishing (see Bar and Kontos, Chapter 4). The program challenged a seasoned musician like Craig to find new ways for his music to inspire movement. He also discovered new purpose and value for his craft through the ways in which the dancers responded to the music he played in the classes. Ruth discovered new meaningful ways to move and Margo found new ways to meaningfully connect with her sister. For Whitney, the program challenged what she thought her clients could do and highlighted their capacity to create and excel. These insights therefore demonstrate ways in which Sharing Dance provided diverse learning opportunities for the entire community and not just for persons living with dementia.

For Craig, Margo, and Whitney, the program also allowed them to get to know people with dementia in a different way and provided a chance to appreciate a person's spirit and different dimensions of their character. This may be one of the most important examples of learning and growth in the program as it challenged assumptions about the capabilities of people living with dementia and older adults in general. Carers, facilitators, and community members had the opportunity to see and learn from the embodied and meaningful self-expression of all participants, and their enjoyment of the program. What they saw challenged their assumptions about what was possible, particularly for participants living with dementia; this resonates with the research of Dupuis et al. (2016) who advocate for the use of art-based programs to challenge dominant assumptions about the actions and abilities of people living with dementia. The narratives above also suggest that Sharing Dance challenged older adults to rethink internalized assumptions about what they were capable of and develop greater confidence in their creative abilities.

Concluding comments

The voices and lived experiences of Sharing Dance featured in this chapter highlight the pleasure that dance can bring to people's lives when it is integrated into their community, places of work, and everyday activities. It can foster moments of joy and social connection as well as opportunities that challenge everyone to learn different things about themselves and others around them. The stories in this chapter corroborate gerontological research on dance, ageing, and well-being that emphasizes the value of everyday dance programs beyond therapeutic interventions. In addition, they show how life-enriching arts-based programs can be for those supporting them. The mutual benefits of arts-based programs are critical to developing sustainable programs that have a lasting impact on the lives of older adults and the attitudes and understanding of younger groups who work and live with them.

References

Bryden, C. (2016). *Nothing about Us, without Us! 20 Years of Dementia Advocacy*. London, UK: Jessica Kingsley Publishers.

Dupuis, S. L., Kontos, P., Mitchell, G., Jonas-Simpson, C., and Gray, J. (2016). Re-claiming citizenship through the arts. *Dementia*, *15*(3), 358–380. https://doi.org/10.1177/1471301216637206

Harris, P. B. (2008). Another wrinkle in the debate about successful aging: The undervalued concept of resilience and the lived experience of dementia. *The International Journal of Aging and Human Development*, *67*(1), 43–61. https://doi.org/10.2190/AG.67.1.c

Herron, R. V., and Rosenberg, M. (2017). "Not there yet": Examining community support from the perspective of people with dementia and their partners in care. *Social Science & Medicine*, *173*, 81–87. https://doi.org/10.1016/j.socscimed.2016.11.041

Houston, S., and McGill, A. (2013). A mixed-methods study into ballet for people living with Parkinson's. *Arts & Health*, *5*(2), 103–119. https://doi.org/10.1080/17533015.2012.745580

Kontos, P., Grigorovich, A., Kosurko, A., Bar, R. J., Herron, R. V., Menec, V. H., and Skinner, M. W. (2021). Dancing with dementia: Exploring the embodied dimensions of creativity and social engagement. *The Gerontologist*, *61*(5), 714–723. https://doi.org/10.1093/geront/gnaa129

Moody, E., and Phinney, A. (2012). A community-engaged art program for older people: Fostering social inclusion. *Canadian Journal on Aging*, *31*(1), 55–64. https://doi.org/10.1017/S0714980811000596

Novek, S., and Wilkinson, H. (2019). Safe and inclusive research practices for qualitative research involving people with dementia: A review of key issues and strategies. *Dementia, 18*(3), 1042–1059. https://doi.org/10.1177/1471301217701274

Sabat, S. (2001). *The Experience of Alzheimer's Disease. Life through a Tangled Veil.* Oxford, UK: Blackwell Publishers.

Schneider, B., and McCoy, L. (2018). Serious leisure, play and the construction of community among older adult square dancers. *Activities, Adaptation, & Aging, 42*(1), 54–68. https://doi.org/10.1080/01924788.2017.1376178

Walsh, K., O'Shea, E., and Scharf, T. (2019). Rural old-age social exclusion: A conceptual framework on mediators of exclusion across the lifecourse. *Ageing and Society, 40*(11), 2311–2337. https://doi.org/10.1017/s0144686x19000606

Wiles, J., Wild, K., Kerse, N., and Allen, R. E. (2012). Resilience from the point of view of older people: 'There's still life beyond a funny knee'. *Social Science & Medicine, 74*(3), 416–424. https://doi.org/10.1016/j.socscimed.2011.11.005

7 Challenging the culture of dementia care through *Sharing Dance*

Pia Kontos and Alisa Grigorovich

Introduction

Dance embraces embodied expression, intersubjectivity, and affect, all of which are essential dimensions of creativity when it comes to dementia (Kontos et al., 2020a, 2020b). As such dance has enormous potential to reduce stigma associated with dementia and to contribute to an important ethical agenda to realize and support dance lives through communal entitlements that promote flourishing for individuals living with dementia in long-term residential care (LTRC) settings. In this chapter, we argue for the need to broaden the therapeutic model of dance to more fully support embodied and creative self-expression by persons living with dementia. In doing so, the chapter contributes an important critical perspective.

Examples of musicality, dance, and other types of creativity in the context of life with dementia abound (Basting, 2009; Kontos and Grigorovich, 2018b, 2019). Yet, examples of such creativity continue to receive vast media attention. Take the example of Henry Dreyer who we first see sitting in a wheelchair, seemingly introverted and disconnected, eyes closed. But then we see a care provider put headphones on Henry attached to an iPod and he becomes animated by the music he hears; when asked what his favourite music was when he was young he responds "Cab Calloway" and breaks into Calloway-style scat talking followed by a soulful rendition of what he later says is his favourite Calloway song – "I'll Be Home for Christmas". This account of Henry features in a documentary film, *Alive Inside* (Rossato-Bennett, 2012) and his performance, which is publicly available on YouTube, has accumulated more than ten million views since it was posted in 2012 (Goodman, 2014). Another more recent video that similarly went viral is of Marta C. González who was a prima ballerina in her younger years (GoodNewsBroadcast, 2020). As Marta sits in her wheelchair, she listens to Tchaikovsky's "Swan Lake" through a pair of headphones, and within moments she begins to reenact the choreographed movements she had once performed on stage. Suddenly there is a flash in her eyes and she becomes Odette, the swan queen at the misty lakeside, with her arms fully extended and peering out to an imagined audience. She is deeply immersed in her performance, using her breath and the music to emphasize her movements which are delicate, graceful, and purposeful.

DOI: 10.4324/9781003260691-7

What is it about these performances that captivates the world? Camic and colleagues note that "'creativity' and 'dementias' are not two words that often find themselves linked" (2018, p. 1) in popular, academic, or empirical narratives. In large part, this is because of the stigma associated with dementia. This stigma is largely based on a particular cultural imaginary (Gilleard and Higgs, 2013), or a collective cultural representation of persons living with dementia as being incapable of purposeful and meaningful communication and interaction. Evidence of this stigma can be seen in media and popular culture representations of dementia as inevitable loss and decline, or a living death (Funk et al., 2020; Latimer, 2018; Mitchell et al., 2013). It is precisely the juxtaposition between the construction of "the hollowed-out person in a state of 'living death'" (Latimer, 2018, p. 837) and the creativity that the videos powerfully convey that explains the world's captivation with them.

Another reason why creativity relative to dementia seems out-of-place is the assumption that creativity is a cognitive "trait" that is associated with the activities and expressions of "gifted" or exceptional individuals (Camic et al., 2018; Kontos and Grigorovich, 2019; Kontos et al., 2018). For example, theories to explain musical intelligence are largely premised upon information-processing models that reduce music perception to a sophisticated stimulus-response system rendered musical by mental manipulation of lower-order stimuli (Bowman, 2004; Kontos, 2014). The implicit assumption is that music perception is cognitive. Without such cognitive intervention – transformation, processing, representation of lower-order stimuli and auditory sense data – the assumption, argues Bowman (2004), is that music would be perceived as little more than "a booming, buzzing confusion" (p. 9). Even with dance, which is a highly embodied form of creative expression, there is a reliance on cognitive science and cognitive and neural processes as an explanatory framework (Batson et al., 2012; Müller et al., 2017; Payne et al., 2016). For this reason, creativity is most commonly associated with originality, genius, and a significant contribution to public life in the sciences, industry, and art – a construction referred to as "Big-C" creativity (Bellass et al., 2019; Kaufman and Beghetto, 2009). There has more recently been a call for the development of a broader view of creativity to account for the everyday and ordinary creativity of "regular" citizens, including persons living with dementia (Bellass et al., 2019; Kontos and Grigorovich, 2018; Kontos et al., 2018). For example, Bellass et al. (2019) have called for more attention to "little-c" creativity, described as "situated" and "mundane" expressions of creativity of persons living with dementia "that take place in familiar everyday situations and spaces" (p. 3) (e.g., creative choices about clothing and accessories).

Building on the discourse of "little-c" creativity, and more broadly participatory arts (de Medeiros and Basting, 2013; Dupuis et al., 2016), scholars have advanced the notion of "co-creativity" (Schmoelz, 2017; Zeilig et al., 2018) to shift the focus from understanding creativity in individualistic terms, to understanding the aesthetic realm of creativity as existing within the social world where creative expressions are constructed and interpreted. With the shift to "little-c" creativity and "co-creativity", there has been a deliberate turn away from conceptualizing creativity solely in cognitive terms, to considering creativity as embodied and relational (Bellass et al., 2019; Zeilig et al., 2018).

Despite these important developments, arts-based programs in LTRC homes (e.g., nursing homes) continue to be implemented as a therapeutic intervention with the aim of managing "behaviours" by reducing neuropsychiatric symptoms associated with dementia (e.g., agitation) and improving cognitive and physical health outcomes (Karkou and Meekums, 2017; Petrovsky et al., 2015). The most common arts-based programs in dementia care include music (Raglio et al., 2008), art therapy (Rusted et al., 2006), and drama (Basting, 2009; Basting et al., 2016). However, dance is receiving increasing attention in dementia care for its potential to support intersubjectivity, and embodied or somatic expression, all of which are essential dimensions of "co-creativity", experience, and care when it comes to dementia (Black et al., 2018; Noice et al., 2014). Yet, just as with other arts-based programs in LTRC settings, dance has largely been restricted in dementia care to its instrumental application as a therapeutic tool to improve "behaviours" and cognitive functioning (Kontos and Grigorovich, 2018a). Given that supporting creativity is not the goal of these programs, they are most often "directed" rather than offered as genuine creative choices for residents, with a model to copy rather than support aesthetic and experimental engagement (e.g., pre-selected songs, paint-by-numbers). They are also typically offered for residents alone rather than as a relationship or community building experience for family, friends, staff, or volunteers to participate in.

In contrast to this, here we explore Sharing Dance Older Adults as an illustrative example of how the arts can support embodied self-expression, co-creativity, social engagement, and inclusion of persons living with dementia in LTRC settings. As we shall argue, it thereby has enormous potential to reduce stigma and to cultivate a new cultural imaginary that embraces the richness of creativity in the context of dementia care.

Playfulness and sociability

In contrast to dance therapy programs that typically emphasize structured repetition of a set of choreographed movements, Sharing Dance uses a narrative approach to dance that encourages participants to express themselves through movement in their own unique way. For example, each movement or inspiration for a movement is introduced in the context of a narrative about common knowledge/experiences and it is up to the participants how they wish to perform or express the movements (for greater details about the structure of the program and its principles see Bar and Dalrymple, Chapter 2 and Table 2.1 in this book, also Kontos et al., 2020b). Take the "Under the Sea" dance as an example. The on-screen instructor (OSI) begins by saying "Let's imagine diving under the sea" and describes the various things one might encounter under water. These include starfish, a shark, sea turtles, snails, and seaweed, and the dance itself involves participants enacting the movements of this ocean life. For example, in the following observation, the OSI describes how seaweed may be the movement's inspiration, and adds that it is up to the individual participants to decide if they are fast or slow moving seaweed, or long or short pieces of seaweed:

> Under the sea I love watching the seaweed as it gracefully floats with the water [gracefully moving her arms up above her head, fluidly bending and curving her elbows, wrists, and fingertips to resemble pieces of seaweed swaying back and forth underwater]. And I'd like you to do any seaweed you would like [fluidly moving one of her curved arms behind her head as she leans her head back, then crossing her arms in front of her before lifting one arm above her head and moving the other across her body; next, increasing the speed of her arm movements and varying the arm movements from above her head to in front of her body, while extending her gaze above her head following her movements]. [Continuing with these movements] You can be your own seaweed. However you feel seaweed moves, that's how I'd like for you to move. Are you a little piece of seaweed [hunched over and with her forearms upright and parallel in front of her, moving her arms and hands closely in front of her in a synchronized wave motion]? Are you maybe a big wild piece of seaweed that's going to catch lots of fish inside of it [chest and arms open, moving her arms in large, swooping, unsynchronized movements in front of her and above her head, along with softly kicking her legs out in front of her]?

The emphasis here on embodied self-expression is apparent even with the narrative nature of the dance. Here we are extending Gray's analysis of the aesthetic and relational practice of clowning (Gray et al., 2019) to highlight how the OSI's words are "rooted in feelings and embodied actions" rather than being a function of intellectual "telling". As such words are an extension of dynamic, embodied interactions that support spontaneous, imaginative, and aesthetic expression (emotional and sensory). The opportunities for self-expression in the "Under the Sea" dance are also inherently playful as they allow participants to try out different ways of moving with the aim of having fun, rather than achieving proficiency with the movements themselves. This is reflected in the following observations that illustrate the diverse ways in which the participants moved as seaweed:

> A participant lifts up both of her hands above her head with a slight bend in her elbows and sways her arms synchronously from side to side, slightly bending her wrists to create a gentle swaying motion. She then brings her arms together above her head and slowly lowers them simultaneously with one arm to each side as if drawing two sides of a large circle. She continues this arm movement but this time alternating between her right and left arms creating a large semi circle with each. She does this with slightly bent wrists, which creates a soft and graceful movement as she lowers her arms.

> He rests his elbows on the arms of his chair and rapidly wiggles his fingers while ever so slightly bending his wrists up and down.
>
> She extends only her left arm in front of her and very briskly flicks her wrist up and down, and then she raises the same arm above her head and wildly swings it in multiple directions.

Another example of the imaginative and playful nature of the dance is the "Railroad" dance which involves digging to lay down the tracks, the wheels of the train moving, and passengers waving. The following observation of this dance captures how participants would let go of what is "real" and become immersed in the narrative of a particular dance, often adding their own style as part of their imaginative offerings:

> The music starts, and, following instruction to wipe the imaginary sweat off their brows, the participants express how hot they are by wiping their arms across their foreheads. One of the participants expresses the heat of the sun slightly differently by raising his hand up over his eyes to create an imaginary visor to shade his eyes.

In the field of dance, this highly individual addition to a dance is referred to as a "signature" or "flourish" that the dancer adds to a pattern of movements associated with a particular dance form that reflects their own personal style (Kraut, 2010; Ophir, 2016). While the scholarship on "flourish/signature" has focused on high-level performance of elite choreographers, dance teachers, and professional dancers (Kraut, 2010; Ophir, 2016), we found that this unique individual offering – what Kraut (2010) refers to as "an assertion of authorship" and "an inscription of identity" (p. 186) – was something the participants with dementia often would make, suggesting that "flourish/signature" should not be confined solely to the domain of the professional artist.

The narrative approach of Sharing Dance not only supports individual playfulness and imagination, but also encourages connectivity/intersubjectivity between participants and their community, including carers, facilitators, and volunteers, all of whom participate in each dance together. Connectivity/intersubjectivity is encouraged by the very nature of the dances of the program, which hinges on participants co-constructing and collaboratively animating the narrative of the dances through gestures, movements, and verbal expressions. This high degree of collaboration hinged on participants being "accomplices" (Gray, 2019) to each other's movements, and collaboratively animating the narrative of the dance. The "Sunrise" dance provides a good example of these embodied and relational dimensions of creativity that effectively transform solitariness of self-expression, to togetherness. Here the OSI invites participants to imagine they are holding the sun in their arms, raising it above their head and releasing it by opening their arms and slowly lowering them to their side. The release of the sun continues with participants being invited to send out sunbeams to each other by stretching out each arm and extending their fingers with an open hand:

"Let's see your sunbeams! Maybe it's to a friend [as the OSI extends her arm and hand out towards the camera]". The following observation powerfully captures how the participants were fully immersed in the narrative and co-constructed the animation of it:

> When I [Research Associate] extend a sunbeam to [a participant living with dementia], he pretends that he has been shot in the chest by grabbing his chest and collapsing his shoulders forward. Another resident participant sends him a sunbeam, and he shoots one back to him with a sound effect – "Zhoom!" He then pretends to eat the sunbeams as they are shot to him, opening his mouth wide and chomping down on the air.

The participants imaginative, brave, and playful extension of their bodies was exemplified by the "signature/flourish" that they often added as in this case of the sunbeams with sound effects, drama, and humour in the form of slapstick style physical spontaneity. This importantly broadens how "signature/flourish" is typically used which is in reference to "movements"; here we extend the concept to include the creative agency of persons living with dementia manifest in both verbal and "kinetic" articulations. By accepting the invitation and imaginatively participating in receiving the sunbeam in the way that he does, the participant inventively disrupts the space and reinvents it as something else with what they do with their bodies. Gray et al. (2019) refer to such disruptions as "playful extending", "experimental gestures", "inventive", and "foolish". By foolish, Gray is drawing on cultural and queer theorist Halberstam's work on failure, creativity, and rethinking seriousness to mean brave, vulnerable experimentation and a willingness to risk being wrong. Foolishness, as Gray writes, is "messy and experimental", uncertain, and risky given the potential for failure. Yet, such risk is precisely what holds the potential to "offer more creative ...more surprising ways of being in the world" (Halberstam quoted in Gray et al., 2019; Grigorovich et al., 2019). Sir Ken Robinson's work on creativity (2006) similarly underscores the importance of risking failure: he argues that if failure is feared, creativity will never be possible. Persons living with dementia themselves reflected on the importance of this:

> [The OSI] told us to not worry about if you get … because I got mixed up a lot … I got lost – so I put my feet and did something [different] than what they were playing… because I've got … Alzheimer's, I get mixed up, but I don't care. I didn't do the same thing that they were doing but I kept going and (shrugs shoulders) I had fun. I really had fun.

> I was always happy to see somebody else making mistakes… Because I was making so many. (Laughter) Yeah, it was reinforcing.

While the residents' participation was deeply engaging, staff, volunteers, and family members were also impacted. Staff described their experience connecting to the participants:

> I liked enjoying the residents' company. I felt like as much as they were connecting with each other, I was connecting with them. It makes me feel so good. Whether we were playing a certain song. When they encouraged you to share – they would put their hand out and touch you. It was human touch. Who doesn't want that?

Volunteers similarly commented on this relational impact: "I really like it when he [the resident] smiled and looked back at me and it made me feel good… It's for him, but you can't help it really, it works both ways, doesn't it?"

In the context of LTRC that is characterized by heavy workloads, low levels of decision-making autonomy, low status, rigid work routines, and insufficient relational care (Grigorovich et al., 2019), it is significant that through volunteers' and staff's engagement with Sharing Dance, they found a meaningful connection with residents. They also experienced an important connection with each other as captured in the following:

> It was great for us as staff members as well to be part of the program and to be engaged with residents that closely. Each week you look forward to it. It was great for us staff – for our morale… it just brought us all a little closer I think.

When considering that the working conditions of LTRC settings are so often described by frontline staff as "stressful", "exhausting", "demoralizing", and as causing "moral distress", reference to the impact that their engagement with the program had on their morale is particularly significant.

Discussion and concluding comments

Our analysis highlights the playful and imaginative nature of the participants' engagement with Sharing Dance, which involved immersion in the narrative dance, suspension of what is real, and adding their own personality to their animation of the dance. This was fully supported by the nature of the program which emphasized creativity by fully embracing the imaginary, rather than focusing on the accuracy of participants following an inventory of movements or sequence of movements. This program stands in stark contrast to other arts-based programs that fail to support the creative potential of persons living with dementia. For example, adult colouring books and "paint by number" are common forms of visual art programs in care homes (Hattori et al., 2011) despite the fact that people living with dementia can engage in unstructured painting with increasing development of skills (Miller and Johansson, 2016).

The communal or shared arts of Sharing Dance makes visible the strengths and capabilities of persons living with dementia and supports new ways of being with each other that enhances social relationships between them and with/between staff, family carers, and volunteers. It allows for the strengths of persons living with dementia to be perceived and valued, which counters assumptions of existential loss with dementia (Grigorovich and Kontos, 2018; Kontos, 2012a; Milne, 2010; Zeilig, 2014). Participation in the program thus has the potential to counter stigma associated with dementia by "highlighting people as individuals with complex lives that exceed the narrow description of diagnosis" (Corrigan, 2007, p. 36), by confirming their vitality through embodied, playful, and imaginative engagement, and thereby supporting social inclusion at the micro-level. As we saw with family, staff, and volunteers, participating with the residents allowed them to enjoy the process of creative expression, which in turn, forged interpersonal relationships and a stronger sense of community. As such, Sharing Dance has the potential to be a powerful catalyst for culture change in dementia care.

In its support of embodied self-expression, co-creativity, and social engagement of persons living with dementia in LTRC settings, Sharing Dance underscores the need for a more inclusive view of creativity – a shift to "little-c" creativity and "co-creativity" and with it a turn away from conceptualizing creativity solely in cognitive terms, to considering creativity as embodied and relational (Bellass et al., 2019; Zeilig et al., 2018). This is consistent with important developments in critical gerontology's subfield of embodiment and dementia, which has vastly expanded and enriched understandings of dementia by placing the body and embodied practices at the centre of explorations of how dementia is represented and/or experienced (Kontos and Martin, 2013). Key to this scholarship has been conceptualizing the body as a generative and creative capacity to perceive and engage with the world and emphasizing the embodied and relational nature of selfhood (Kontos, 2012a, 2012b; Kontos and Martin, 2013). Recent developments have expanded this discourse even further by establishing links between embodiment, relationality, and the importance of more inclusively granting citizenship and human rights entitlements to persons living with dementia in LTRC on the grounds, including access to opportunities to be creative to the fullest extent possible (Kontos et al., 2020a). The implications of this for our analysis is that the body as a source of agency is not only fundamental to creativity but also must be recognized as fundamental to the human condition. As such, it is imperative that all forms of creativity (i.e., Big-C, little-c) is supported through public policy and the (re)distribution of social resources to ensure that states and their institutions act in ways that do not produce or perpetuate barriers to equality and equal enjoyment of the rights in this area (Kontos et al. 2018a; Kontos and Grigorovich 2018b, 2018c). Moreover, to ensure realization of these rights, in addition to removing barriers, this would require the mobilization of structures and resources to nurture and facilitate opportunities for the creativity of persons living with dementia in their everyday life.

In the case of Sharing Dance, participants are presently supported to participate with funding from multiple levels of government and generous philanthropists, as well as from earned revenue from subscription-based services and business relationships with participating retirement living and LTRC (see Kosurko, Arminen, and Stevanovic, Chapter 8). These resources are integral to the sustainability of the program. Yet, the argument that creativity is supported through a matrix of human rights requires more than occasional and precarious public funding of arts-based programs and reliance on charity; we need creativity to be intentionally and securely supported with dedicated and stable arts-based program funding. Yet, we are advocating even more than this. For creativity to be fully supported, it cannot be confined to structured social programs. This is not to say that Sharing Dance does not inspire participants to dance outside of the structured sessions; indeed there are many anecdotal examples of this (see Bar, Dunphy, Herron, Snider, Strachan, and Wingrove, Chapter 6). However, and in the context of LTRC, supporting creativity in everyday life requires that we diffuse the responsibility for enabling the creative and emotional enrichment of the LTRC environment. This would ensure that residents' creativity is supported and nurtured in all aspects of institutionalized life and not exclusively through arts and leisure programming (Genoe and Dupuis, 2014; Moss and O'Neill, 201). Such initiatives require fostering a social ecology of caring, corporeal-ethical spaces that value and support embodiment and relationality so that they become part of the moral fabric of everyday life in LTRC settings (Kontos and Grigorovich, 2018a). It entails an "aesthetic approach to care" (Gray et al., 2019), central to which is valuing bravery, a willingness to fail, and the infinite possibilities that risk in the creative sense of "daring and adventure" would afford (Kontos, 2022). This is in keeping with urgent calls for dementia care to sustain human flourishing by supporting embodied and relational capabilities (Grigorovich et al., 2019), central to which is the creativity and imagination (Kontos et al., 2020b). Given that critical approaches to understanding creativity in the context of dementia is a nascent area of research, there is much work to be done to learn from these experiences and to reimagine dementia care to realize the full creative potential of persons living with dementia in LTRC settings.

References

Basting, A. (2009). *Forget Memory*. Baltimore, MD: Johns Hopkins University Press.

Basting, A., Towey, M., and Rose, E. (2016). *The Penelope Project: An Arts-Based Odyssey to Change Elder Care*. Iowa City: University of Iowa Press.

Batson, G., Quin, E., and Wilson, M. (2012). Integrating somatics and science. *Journal of Dance & Somatic Practices*, *3*(1–2), 183–193. https://doi.org/10.1386/jdsp.3.1-2.183_1

Bellass, S., Balmer, A., May, V., Keady, J., Buse, C., Capstick, A., ... Hodgson, J. (2019). Broadening the debate on creativity and dementia: A critical approach. *Dementia: The International Journal of Social Research and Practice*, *18*(7–8), 2799–2820. https://doi.org/10.1177/1471301218760906

Black, H. K., Rubinstein, R. L., Frankowski, A. C., Hrybyk, G., Nemec, M., and Tucker, G. G. (2018). Identity, semiotics, and use of symbols in adult day services. *The Gerontologist*, *58*(4), 730–738. https://doi.org/10.1093/geront/gnx074

Bowman, W. D. (2004). Cognition and the body: Perspectives from music education. In L. Bresler (Ed.), *Knowing Bodies, Moving Minds: Toward Embodied Teaching and Learning* (pp. 29–50). Netherlands: Kluwer Academic Press.

Camic, P. M., Crutch, S. J., Murphy, C., Firth, N. C., Harding, E., Harrison, C. R., … Zeilig, H. (2018). Conceptualising and understanding artistic creativity in the dementias: Interdisciplinary approaches to research and practise. *Frontiers in Psychology*, *9*, 1–12. https://doi.org/10.3389/fpsyg.2018.01842

Corrigan, P. W. (2007). How clinical diagnosis might exacerbate the stigma of mental illness. *Social Work*, *52*(1), 31–39. https://doi.org/10.1093/sw/52.1.31

de Medeiros, K., and Basting, A. (2013). "Shall I compare thee to a dose of Donepezil?": Cultural arts interventions in dementia care research. *The Gerontologist*, *54*(3), 344–353. https://doi.org/10.1093/geront/gnt055

Dupuis, S., Kontos, P., Mitchell, G., Jonas-Simpson, C., and Gray, J. (2016). Re-claiming citizenship through the arts. *Dementia: The International Journal of Social Research and Practice*, *15*(3), 358–380. https://doi.org/10.1177/1471301216637206

Funk, L. M., Herron, R. V., Spencer, D., and Thomas, S. L. (2020). Aggression and older adults: News media coverage across care settings and relationships. *Canadian Journal on Aging*. https://doi.org/10.1017/s0714980820000197

Genoe, M. R., and Dupuis, S. L. (2014). The role of leisure within the dementia context. *Dementia*, *13*(1), 33–58.

Gilleard, C., and Higgs, P. (2013). The fourth age and the concept of a 'social imaginary': A theoretical excursus. *Journal of Aging Studies*, *27*(4), 368–376. https://doi.org/10.1016/j.jaging.2013.08.004

Goodman, A. (2014). Alive Inside: How the magic of music proves therapeutic for patients with Alzheimer's and dementia. https://www.democracynow.org/2014/1/22/alive_inside_how_the_magic_of

GoodNewsBroadcast. (2020, November 11). *Marta Gonzalez former Prima Ballerina with Alzheimer's* [Video]. YouTube. https://www.youtube.com/watch?v=hsLLXY_wZYI

Gray, J. (2019). Working within an aesthetic of relationality: Theoretical considerations of embodiment, imagination and foolishness as part of theatre making about dementia. *Research in Drama Education: The Journal of Applied Theatre and Performance*, *24*(1), 6–22. https://doi.org/10.1080/13569783.2018.1535270

Gray, J., Donnelly, H., and Gibson, B. E. (2019). Seriously foolish and foolishly serious: The art and practice of clowning in children's rehabilitation. *Journal of Medical Humanities*, *42*(3), 453–469. https://doi.org/10.1007/s10912-019-09570-0

Grigorovich, A., and Kontos, P. (2018). Advancing an ethic of embodied relational sexuality to guide decision-making in dementia care. *The Gerontologist*, *58*(2), 219–225. https://doi.org/10.1093/geront/gnw137

Grigorovich, A., Kontos, P., and Kontos, A. P. (2019). The "violent resident": A critical exploration of the ethics of resident-to-resident aggression. *Journal of Bioethical Inquiry*, *16*(2), 173–183. https://doi.org/10.1007/s11673-019-09898-1

Hattori, H., Hattori, C., Hokao, C., Mizushima, K., and Mase, T. (2011). Controlled study on the cognitive and psychological effect of coloring and drawing in mild Alzheimer's disease patients. *Geriatrics & Gerontology International*, *11*(4), 431–437. https://doi.org/10.1111/j.1447-0594.2011.00698.x

Karkou, V., and Meekums, B. (2017). Dance movement therapy for dementia. *Cochrane Database of Systematic Reviews*, *2017*(2), 1–29. https://doi.org/10.1002/14651858.cd011022

Kaufman, J. C., and Beghetto, R. A. (2009). Beyond big and little: The four c model of creativity. *Review of General Psychology*, *13*(1), 1–12. https://doi.org/10.1037/a0013688

Kontos, P. (2012a). Alzheimer expressions or expressions despite Alzheimer's?: Philosophical reflections on selfhood and embodiment. *Occasion: Interdisciplinary*

Studies in the Humanities, *4*, 1–12. http://arcade.stanford.edu/sites/default/files/article_pdfs/OCCASION_v04_Kontos_053112_0.pdf

Kontos, P. (2012b). Rethinking sociability in long-term care: An embodied dimension of selfhood. *Dementia: The International Journal of Social Research and Practice*, *11*(3), 329–346. https://doi.org/10.1177/1471301211421073

Kontos, P. (2014). Musical embodiment, selfhood, and dementia. In L. C. Hydén, J. Brockmeier, and H. Lindemann (Eds.), *Beyond Loss* (pp. 107–119). New York: Oxford University Press.

Kontos, P. (2022). Valuing risk in residential long-term care: Setting an important ethical standard for supporting and nurturing human flourishing. In K. de Medeiros, T. Cole and M. Goldman (Eds.), *Critical Humanities and Ageing: Forging Interdisciplinary Dialogues*. Abingdon, UK: Routledge

Kontos, P., and Grigorovich, A. (2018a). Integrating citizenship, embodiment, and relationality: Towards a reconceptualization of dance and dementia in long-term care. *The Journal of Law, Medicine & Ethics*, *46*(3), 717–723. https://doi.org/10.1177/1073110518804233

Kontos, P., and Grigorovich, A. (2018b). Rethinking musicality in dementia as embodied and relational. *Journal of Aging Studies (Special Issue: Aging, body and society. Key themes, critical perspectives, Eds., Martin, W. & Twigg, J.)*, *45*, 39–48. https://doi.org/10.1016/j.jaging.2018.01.006

Kontos, P., and Grigorovich, A. (2018c). Dancing with dementia: Citizenship, embodiment and everyday life in the context of long-term care. In S. Katz (Ed.), *Ageing in Everyday Life: Materialities and Embodiments* (pp. 163–180). Bristol, UK: Bristol University Press.

Kontos, P., and Grigorovich, A. (2019). A critical narrative on late life creativity and dementia: Integrating citizenship, embodiment, and relationality. In D. Amigoni and G. McMullan (Eds.), *Creativity in Later Life: Beyond Late Style* (pp. 167–181). Oxon, UK: Policy Press.

Kontos, P., and Martin, W. (2013). Embodiment and dementia: Exploring critical narratives of selfhood, surveillance, and dementia care. *Dementia: The International Journal of Social Research and Practice*, *12*(3), 288–302. https://doi.org/10.1177/1471301213479787

Kontos, P., Grigorovich, A., Kontos, A. P., and Miller, K. L. (2018). Exploring relational citizenship at the intersection of creativity and dementia. In H. P. Zimmerman (Ed.), *Kulturen der Sorge: Wie unsere Gesellschaft ein Leben mit Demenz ermöglichen kann* (pp. 543–560). Zürich: Campus-Verlag.

Kontos, P., Grigorovich, A., and Colobong, R. (2020a). Towards a critical understanding of creativity and dementia: New directions for practice change. *International Practice Development Journal*, *10*(Art 3). https://doi.org/10.19043/ipdj.10Suppl.003

Kontos, P., Grigorovich, A., Kosurko, A., Bar, R. J., Herron, R. V., Menec, V., and Skinner, M. W. (2020b). Dancing with dementia: Exploring the embodied dimensions of creativity and social engagement. *The Gerontologist*, *61*(5), 714–723. https://doi.org/10.1093/geront/gnaa129

Kraut, A. (2010). "Stealing steps" and signature moves: Embodied theories of dance as intellectual property. *Theatre Journal*, *62*(2), 173–189. https://doi.org/10.1353/tj.0.0357

Latimer, J. (2018). Repelling neoliberal world-making? How the ageing–dementia relation is reassembling the social. *The Sociological Review Monographs*, *66*(4), 832–856. https://doi.org/10.1177/0038026118777422

Miller, E., and Johansson, B. (2016). Capability to paint and Alzheimer's disease: Relationship to disease stages and instructions. *Sage Open*, *6*(1), 1–13. https://doi.org/10.1177/2158244016631799

Milne, A. (2010). The 'D' word: Reflections on the relationship between stigma, discrimination and dementia. *Journal of Mental Health*, *19*(3), 227–233. https://doi.org/10.3109/09638231003728166

Mitchell, G., Dupuis, S., and Kontos, P. (2013). Dementia discourse: From imposed suffering to knowing other-wise. *Journal of Applied Hermeneutics*, 1–19. https://journalhosting.ucalgary.ca/index.php/jah/article/view/53220

Moss, H., and O'Neill, D. (2014). Aesthetic deprivation in clinical settings. *The Lancet*, *383*(9922), 1032–1033.

Müller, P., Rehfeld, K., Schmicker, M., Hökelmann, A., Dordevic, M., Lessmann, V., … Müller, N. G. (2017). Evolution of neuroplasticity in response to physical activity in old age: The case for dancing. *Frontiers in Aging Neuroscience*, *9*, 1–8. http://www.ncbi.nlm.nih.gov/pmc/articles/PMC5348543/. https://doi.org/10.3389/fnagi.2017.00056

Noice, T., Noice, H., and Kramer, A. F. (2014). Participatory arts for older adults: A review of benefits and challenges. *The Gerontologist*, *54*(5), 741–753. https://doi.org/10.1093/geront/gnt138

Ophir, H. (2016). The signature of the moving body: Agency and embodied education ideologies of dance teachers. *Anthropology, Education Quarterly*, *47*(2), 186–202. https://doi.org/10.1111/aeq.12148

Payne, H., Warnecke, T., Karkou, V., and Westland, G. (2016). A comparative analysis of body psychotherapy and dance movement psychotherapy from a European perspective. *Body, Movement and Dance in Psychotherapy*, *11*(2–3), 144–166. https://doi.org/10.1080/17432979.2016.1165291

Petrovsky, D., Cacchione, P. Z., and George, M. (2015). Review of the effect of music interventions on symptoms of anxiety and depression in older adults with mild dementia. *International Psychogeriatrics*, *27*(10), 1661–1670. https://doi.org/10.1017/s1041610215000393

Raglio, A., Bellelli, G., Traficante, D., Gianotti, M., Ubezio, M. C., Villani, D., and Trabucchi, M. (2008). Efficacy of music therapy in the treatment of behavioral and psychiatric symptoms of dementia. *Alzheimer Disease and Associated Disorders*, *22*(2), 158–162. https://doi.org/10.1097/wad.0b013e3181630b6f

Rossato-Bennett, M. (Writer). (2012). *Alive Inside: A Story of Music and Memory* [Film]. Projector Media.

Rusted, J., Sheppard, L., and Waller, D. (2006). A multi-centre randomized control group trial on the use of art therapy for older people with dementia. *Group Analysis*, *39*(4), 517–536. https://doi.org/10.1177/0533316406071447

Schmoelz, A. (2017). On co-creativity in playful classroom activities. *Creativity: Theories–Research-Applications*, *4*(1), 25–64. https://doi.org/10.1515/ctra-2017-0002

Zeilig, H. (2014). Dementia as a cultural metaphor. *The Gerontologist*, *54*(2), 258–267. https://doi.org/10.1007/978-3-030-22009-9_263

Zeilig, H., West, J., and van der Byl Williams, M. (2018). Co-creativity: Possibilities for using the arts with people with a dementia. *Quality in Ageing and Older Adults*, *19*(2), 135–145. https://doi.org/10.1108/qaoa-02-2018-0008

8 Technological glitches and creative interactions in *Sharing Dance*

An Kosurko, Ilkka Arminen, and Melisa Stevanovic

Introduction

Screens, the internet, and technology are typically not what you might think of when it comes to dancing or engaging with the arts. For the Sharing Dance Older Adults program developed by Baycrest and Canada's National Ballet School (NBS) however, technology plays a critical role. To provide wider geographical access to a program designed to be accessible to older people with cognitive and physical limitations, information communication technology (ICT) is the key – and a few glitches are to be expected when making it work. As outlined in Chapter 2 (Bar and Dalrymple) in this book, the mechanisms put in place to support the remote delivery of the dance program affected how dance teachers would tailor instructions to participants from afar. The aim of this chapter is to explore the role of ICT in expanding remote access to those instructions and the technological limitations that influenced experience of the digital Sharing Dance program. To further this exploration, we will introduce an additional step inspired by the research project as the digital expansion extends across international borders to Finland. After providing some background on ICT in the context of social inclusion for rural older adults, we will talk about how participants experienced the use of technology to access the online dance classes. We will discuss what participants said about their experiences and what we observed as they interacted in response to technical difficulties during the pilot studies. We will then pose a new analytic angle to the study that builds on this work in an international context, through a close examination of these same types of interactions. This chapter is about the process of research and looking at similar phenomena in different ways, in this case the technical difficulties commonly referred to as glitches.

Technology in the context of social inclusion for rural older adults

Technology is taken for granted as the great problem solver of our time. The COVID pandemic has shown us how dependent upon technology we can become to connect across distances and to be productive through time zones. Yet, in terms of social inclusion, access to technology is not always equal.

DOI: 10.4324/9781003260691-8

Many are excluded from services and programs as they are increasingly mediated by technology. In rural areas in Canada, higher proportions of older adults live where there is a double digital divide – where limited ICT infrastructure impedes the remote delivery of programs and services – and where some people lack familiarity in use of the technology (O'Connell et al., 2018).

Social inclusion has been prioritized by many agencies from global to local levels. It is a broad concept encompassing many themes that can each be looked at from different perspectives and at different levels. For the purpose of our study on enhancing social inclusion, we utilized and adapted a framework developed by Walsh et al. (2020) that makes sense of the relational nature of multiple levels of social exclusion for rural older adults in interlinked domains. In this chapter, we focus specifically on the role of technology in Sharing Dance and the relevant interlinked domains of social relations and access to services and resources. We explore interpersonal dynamics in participant experiences of the on-screen dance instruction and discuss institutional and community capacity to access the digital resource. Technology in this context can be seen as a mediating force that can both alleviate and exacerbate barriers to social inclusion.

One of the important lessons that we learned as part of our broader study was that different types of technology affected social inclusion in different ways and that remote delivery is a complex process involving multiple stakeholders (see Kosurko et al., 2022). As a mediating factor, ICT improved access to an important service by connecting older adults in rural areas to a dance program, while at the same time, it presented new challenges in areas with differing capacities to operate the technology such as internet infrastructure and individual comfort levels with operating ICT equipment (see Kosurko et al., 2020a).

Other previous studies have shown that ICT can be used to address issues of older adult social exclusion such as social isolation for older people (Chen and Schulz, 2016) including people living with dementia (Pinto-Bruno et al., 2017), and those who live in rural areas who may not use technology often (Warburton et al., 2013). However, challenges have been identified for organizations to be able to access programs and services delivered by ICT including training of staff (Van Der Heide et al., 2012). This chapter explores how the use of digital technology influenced participants' experiences in accessing an arts-based program to enhance social inclusion.

Examining the role of technology in Sharing Dance

To explore the role of technology in the Sharing Dance program as it relates to enhancing social inclusion, we look at data collected from 2017 to 2019 as part of the *Improving Social Inclusion for People with Dementia and Carers through Sharing Dance* project when the eight-week Sharing Dance program pilots were delivered in community and long-term residential care (LTRC) settings in 12 non-metropolitan settings in two Canadian provinces: Ontario and Manitoba (see Skinner and Bar, Chapter 3). Interviews were conducted for participants' personal reflections of their experiences with the dance program and how it was delivered. Observations focused on embodied interactions and expressions during weekly sessions, recorded in field notes structured by a semi-structured guide that developed from pilot to pilot. Field notes included reflective and descriptive accounts of settings and participants; and interactions between participants and on-screen instructors (OSI) during the program.

Developed by Baycrest and NBS, Sharing Dance aims to make dance accessible to older people with a range of physical and cognitive abilities, including people living with dementia (see Bar and Dalrymple, Chapter 2). Originally developed to be delivered in-person, the program was subsequently adapted for remote delivery through video streaming of on-screen instruction of dances with seated or standing options. On-site facilitators with experience leading older adults in physical activity were identified locally for each site (in both community and LTRC settings, for example, the recreation director or yoga instructor in the community setting or the recreational therapist in an LTRC setting) and supported the delivery of the program. Research participants included older people, people living with dementia, and family carers (ranging from 66 to 96 years old); administrators and staff in both community and institutional settings; facilitators and volunteers supporting participants. There were a total of 289 participants in the three phases in both regions. Technology for the pilot study was provided where necessary by research partners and with community foundation funding that allowed for the purchase of large-screen smart TV monitors and laptops and signal boosters (rocket hubs) to enhance Wi-Fi connectivity. Installation and technical support were provided by the research partner and program provider (see Skinner and Bar, Chapter 3 for a full description of the research design and methods).

Glitches, glitching, and the struggles with technology in the experiences of Sharing Dance

What we learned about using technology for the Sharing Dance program came from participant experiences and perceptions as well as our research team observations and reflections in the field. We examined what participants said about the delivery of the program in response to questions in interviews and focus groups, as well as the field notes from observed dance classes during the eight-week sessions. Participants included older adults as well as their formal and informal carers, (staff, family, volunteers) and the facilitators who were key players in the delivery of the program. During interviews and focus groups, the subject of technology came up in response to questions such as: How did the mode of delivery influence the experience (i.e., specifying live vs. pre-recorded stream, on-screen instruction, etc.)? What aspects of facilitating the sessions did you feel were most challenging (i.e., for facilitators)? Can you talk about a particular moment in a session or an experience during the program that you did not enjoy? What did you like least about this program/your role?

In the transcripts of interviews and focus groups, as well as field notes, diaries, and research team meeting notes, technical difficulties emerged as a prominent theme, reported as a challenge in the early Peterborough pilots (Warrener et al., 2017). In all subsequent pilots, we assigned technology as a thematic node using NVivo12 software in the qualitative data analysis that examined the text in all of the documents. Technology was consistently reported as a challenge in all pilot reports. We applied a word frequency query to the technology node in data from the final pilots for each region and found that the most common words (top four out of the top 15 words excluding "screen" and "group") used to describe the mode of delivery involving technology were: "glitches," "glitching," "struggled," and "difficulties" (see Figures 8.1 and 8.2 for graphic depiction of word frequency).

Using the NVivo12 software, we determined that glitches were a prominent feature of the described experiences of technology during the research project. Looking more closely at examples of when and how the word "glitch" was uttered, we identified varying perspectives on and approaches to the influence of technology on the Sharing Dance program. In the following section, we look first at what was said about the technical glitches by research participants when they were asked directly in interviews about their

Figure 8.1 Word frequency in mode of delivery/technology node for Peterborough pilot study.

Figure 8.2 Word frequency in mode of delivery/technology node for Brandon pilot study.

experiences. We then look at what the program participants did in situ – how they responded to situations when the described glitches were observed.

Dancers versus facilitators: "It's no big deal," "but technical difficulties are the worst"

Many of the Sharing Dance participants described how the OSI at NBS taught the remote classes in such a way that they felt "as if they were in the same room," "talking directly to them." It was only when the technology would stutter or "glitch" that this illusion was interrupted and the experience in some cases turned into frustration. Looking at examples of what participants had to say about these glitches gave us insights about the different perspectives involved in the use of technology to deliver the program remotely. Many participants, including the older people interviewed about their participation in a community centre below, described their experience of the program delivery as "good," with the technical glitches or interruptions being a minor disturbance in the program:

> It was ok, but I did notice occasionally there was some interruption in the video – that happens.
>
> It was good. The video cut out a few times but life goes on– it's a few seconds really. It's not a big deal.

Each of these participants described how the interruptions of the program were acceptable as part of the video delivery and a minor disruption in their experience. For facilitators, however, technology emerged more often as a theme when they talked about what they found most challenging or what they liked least about the program or their role, as illustrated by the interview responses of three facilitators below:

> Trying to keep the computer running, having glitches – technical difficulties are the worst things.
>
> When the technology was hiccupping, I could see the frustration on people's faces. I felt helpless and that made it disappointing.
>
> And the only other thing that made it difficult was technology, but there's nothing we can do about that. Obviously. Every week – other than the last two – every week we had glitching moments when it would stop and we would still be singing and we would just kind of wait.

In their role to deliver the program to people in the room, facilitators faced frustration directly from their co-present participants as they waited through internet delays on repeated occasions. Describing feelings of helplessness and saying that "there was nothing we could do about it" indicates their dependence upon the smooth operation of the technology to make the program work. While participants indicated that the technical difficulties were no big deal, the facilitators voiced experiences of stress that were more negative concerning the technical aspects of accessing the program remotely.

These frustrations were also described by facilitators with reference to setting up the program:

> Just the set up – I just find that we are not properly outfitted to do a program in that manner, so it's the fact of having to pull the TV out plug things in, run this cord, get speakers out, get online, change this setting, then not being able to get online, then waiting for [the researcher's] arrival in order to be able to use the [Wi-Fi rocket] hub - that was probably my least [favourite aspect].

Many facilitators raised concerns about the steps and equipment involved with operating the program. In the above quote, the facilitator talks about how the research associate provided additional equipment, a Wi-Fi rocket hub, in order to boost the internet signal for the smooth streaming of the video streams. This also points to the dependence on the resources provided by the research project that made access to the program possible (see Menec, Skinner, and Kosurko, Chapter 9 for a deeper discussion on the research influence).

In addition to setting up equipment, facilitators would access the online program by logging in to the web-based platform on the computer. This was a challenge for some facilitators – and others who would have to access the program. One volunteer who filled in for a facilitator described the experience of logging-in her interview:

> That was stressful – I'm sure if I did it a couple of times and actually it was my stuff and I would have passwords written down and practicing it ten times before hand – just the logging in seemed to … always be stressful there.

In the situation described above, the process of logging in was associated with stress because the volunteer was using equipment that was not their "stuff" and in that sense was not an expert familiar with the local technological terrain. Considerations would need to be made for non-expert users to function in place of the expert who possesses deeper understanding of the material setting such as where passwords were recorded (Arminen and Poikus, 2009). This adds to considerations for a variety of users to be able to access the program within the same facilities as well as from different locations, emphasizing the dependence upon individual facilitators to run the program within their "own" facilities.

Our analysis of what was said about the technical glitches indicates that participants in the dance program were not as stressfully affected by the technical difficulties as the facilitators running the program. Understanding facilitators' experiences is salient in recognizing the smooth operation of the program as a co-participant leading others engaged in the program. Facilitators' insights were helpful in the development of the program to define the role of facilitators and requirements for their training and support in accessing and operating remote programs smoothly. As the technology develops in response to these "user experiences," the support for facilitators would also need to adapt and evolve. (For more about the development of delivery, see Kosurko et al., 2021). More support is necessary for various facilitators as important end-users of the program in different settings.

Keep going, even if we glitch! Responses to technical difficulties

In addition to listening to what participants and facilitators said about the technology involved with the remote delivery of the program, we also examined what they did in response to technological glitches while they were participating in the dance. In field notes, we could reference descriptions of participants' interactions during the glitches. In one example captured in a field note below, the internet connection stopped, and the group was unable to continue the program. The facilitator, staff, and participants worked together to remember the final two dance routines from the program that they had learned in the previous weeks:

> The facilitator goes to the computer and attempts to adjust the internet connection and still the program won't restart. The facilitator sits down and says, "I'm going to finish it," and with the support of staff, thinks through to remember the sequence of the moves from previous weeks and then leads the class in the baseball sequence. The facilitator gets everyone singing "Take Me Out to the Ballgame" and they run the routine three times all together. All in the group follow along. … then they lead the group through a singalong with swaying arms to "You are my sunshine," which the group follows along with. When the song finishes, they announce refreshments and all applaud.

The group in this situation was able to complete the program from their collective memory, given that it was their third week in the program. Rather than giving up and ending the session, the facilitator made the decision to "finish it," and with reminders as to what was next from co-participants, they were able to collaborate to complete the dance as they had learned it. This is of interest because it demonstrates participants' engagement with the program remotely and how they were able to realize the Sharing Dance program even when the technology failed to deliver it in "real time." The ability of the group to work together to remember the dance they had been taught twice previously also speaks to the strength of the program to engage communities in dance from beyond the dance teachers' presence in the room. While the completion of this particular dance class may have been somewhat of an interpretation, the participants shared a dance they had learned as it was delivered remotely by NBS. This shared endeavour by participants would not have been possible without the facilitator having assumed the role of the leader, which not all facilitators might feel comfortable doing.

In addition to trying to finish the dances as they were taught, participants responded to technical glitches in other ways. In the example below, from field notes in the Brandon pilot, participant Seth freezes his body in suspended motion and makes a joke:

> There is a pause with the Wi-Fi – the rocket hub icon is flashing as connectivity is limited. Seth makes a joke, his foot is suspended in the air. "This is going to be awkward…"

The next example is similar to the situation above, where the program stops and everyone freezes in response. In this field note, the facilitator likens it to a children's game of statue, where players have to not move in order to stay in the game:

> "Oh dear," says one as the computer glitches. "Then we should scratch," says the facilitator – referring to the movements that should follow the mosquito bite. She also jokes that it's like playing statue as a kid – remember?

Here the facilitator suggests that the dancers should pretend to "scratch," in keeping with the narrative of the cottage scene that they were just performing, in a continuation of the engagement with the program narrative while it was glitching. In each of the examples of freezing moments above, there is a sense that the program will start again, and participants keep their bodies suspended in wait. Alternatively, participants were also observed dancing through the glitches, understanding that they would continue to happen. The field note below describes a moment where the glitching is repetitive:

> All are moving along in sync with the instructor, the facilitator and each other, moving their legs forward and back, brushing the floor with their feet on the way out for a "shuffle – and back," placing their foot back under their chair, then switching feet. Nick smiles, looking up to the screen and moves his feet in and out, a smile still on his face. "Out – out – in," says the on-screen instructor, "Out out," and then the computer glitches and the facilitator says, "In!" to finish the sequence… The computer starts and stops a few times. "Keep on going," says the facilitator as she continues the movements, "even if we glitch."

In this example, the facilitator again continues to lead the program where the OSI left off, by completing the sentence with "in!" Then the facilitator explicitly instructs the group to keep going with the movements following her demonstration. This is another example of how the participants were able to remain engaged in the program regardless of the technology glitches through the participation of the dancers as led by the facilitator.

There were two ways that we accessed our understanding of the use of the word "glitches" and moments of technical difficulties from varying perspectives: we observed situations and recorded with field notes what we saw; then we asked people what they experienced and recorded what they said. From what people said, we found discrepancies in attitudes and opinions about the influence of technology on their experiences. From a participant's perspective, technical difficulties were "no big deal," but from the perspective of the facilitator responsible for implementing the program, technology involved moments of stress and frustration. This gave us insights and allowed us to make recommendations about the different roles involved in the implementation of the remote delivery of the program. From what we observed, we could identify different ways that participants (including facilitators) responded in situations when glitches happened. They collaborated to finish the sequence as they had learned it; they paused and waited for the program to resume; and they improvised an imaginary continuation of the scenario provided by the program.

New horizons to explore social inclusion mediated by technology

As the role of technology in the development of Sharing Dance became clearer during the data collection period, new research questions developed seeking to understand how technology mediates the social connectedness of rural older adults in remotely delivered programs like Sharing Dance. The research associate (and lead co-author of this chapter) developed a Ph.D. project to explore this phenomenon in an international context that would take the Sharing Dance program and research to Finland (see Kosurko et al., 2021).

This new research project looks at social connectedness in the context of the dance sessions with a group outside of Canada. It focuses on the micro level of interpersonal social interaction, in order to observe how participants achieve social connectedness while interacting in an online dance sequence. How technology mediates this interaction can be observed in the procedures people use to together make sense of what is going on in situations where technology is used. The procedures that people use to make sense of their everyday circumstances are referred to in social research as ethnomethodology and a tool that is often used to look closely at this phenomenon is conversation analysis (Heritage, 1984). When conversation analysis is used to look at how people make sense of situations together, it is called ethnomethodological conversation analysis (EMCA). Emphasis is placed on observable actions in sequence, rather than personal introspection as the source of data (Moore, 2013). This research method builds up the details of sequences of interaction as they are observed, to analyse interactions as they occur as well as how they are described in participants' personal reflections of their experiences. In the context of the Sharing Dance program and social inclusion, we can detail within the sequences of interaction where technology affords or constrains participants to orient to the dance program in parallel, co-present activity (Arminen et al., 2016).

Building on the work in the previous section that used a qualitative, thematic analysis of participants' descriptions, a detailed analysis of the interaction mediated by technology can be made using EMCA tools by drawing on "the resources of language, the body, the environment of the interaction, and position in the interaction fashioned into conformations designed to be, and to be recognizable by recipients as, particular actions" (Schegloff, 2007, p. xiv.). The process of EMCA starts with an initial noticing of a moment of interest within an interaction. Then actions are identified in turn taking sequences and collections of these are compiled to build a claim and support it with evidence. These observable outcomes are crucial to support analytic claims, demonstrating how participants respond to actions in ways that we can observe. This approach can be used to gain insights into how relationships are built in social actions, how bodies relate to each other or their environments, interactions between body, talk, and the material environment, turn taking, who orients to whom and how those orientations change. Our special interest in applying this method to Sharing Dance is in how social connectedness is made visible and maintained in digitally mediated dance.

To illustrate the EMCA method, we show a preliminary example from the international pilot in Finland of how participants respond to a technical glitch in the Sharing Dance video in Figure 8.3. We will demonstrate a sequence of turns in a series of images with transcriptions to detail the order of actions. Commentary is provided in the text bubbles. The sequence will show how Sharing Dance participants, Sirpa and Leena, respond to the OSI and music in the video. When the program stops due to a glitch in the technology on Line 7, Sirpa freezes in suspended motion and Leena looks to the facilitator. Two seconds later, when the video resumes, Sirpa joins the dance from where it – and she – left off, with renewed vigour, laughing.

In Figure 8.3, participants in the picture frames Sirpa (left) and Leena (right) are sitting so that they can see the TV screen. The inset figure of the OSI is the on-screen instructor. Off-screen, there is a facilitator (F), who is also following the movements on the screen.

Sirpa freezes in position (Line 8) and remains suspended for almost two seconds until the music starts again. Her action of the dance in progress is interrupted by the technological glitch. Lerner and Raymond (2021) would refer to this glitch as a visible source of body trouble, to which Sirpa responds by adjusting her action (freezing). In a sense, it is the technology's turn to make a move in the interaction. As a computer system and machine, there is no intention behind the technology in creating this interruption (Suchman, 1987), but it sets up a response from the participant in the dance program nonetheless. The frozen pause raises a question about how this participant (Sirpa) orients to the technology in relation to co-participants – that may be answered by what happens next in the order of the sequence.

After Sirpa freezes, Leena looks to the facilitator on Line 9, then turns their head to the screen, acknowledging first an awareness of co-participants in the room and then orienting to the screen as the source of the trouble. The facilitator's comment, "Desolana" (Spanish for "nothingness") on Line 13 is an example of how participants use humour to manage disruptions, in an attempt to maintain focus on the collective task. In this instance, the facilitator is using the technological glitch as a potential relational resource, rather than something that is blocking interaction (Rintel, 2013). However, Sirpa neither responds to the facilitator's comment nor looks away from the screen. She orients to and prioritizes the manual activity of the dance sequence, frozen until the technology resumes the dance instruction. Her embodied conduct (freezing) demonstrates that her priority orientation is with the manual activity of the on-screen dance choreography, but not the interactional "trouble" of her co-present participants (Kamunen, 2019). In the next moment however, Sirpa shows that she is also still aware of others in the room as capable of seeing her performance of freezing and moving again, made evident by her enthusiastic laughter on Line 16 when the program resumes. These moments considered in sequence as unfolding in relation to each other denote a social relevance: Leena reacts to Sirpa's freezing by looking to the facilitator for an explanation, which prompts a joke from the facilitator while

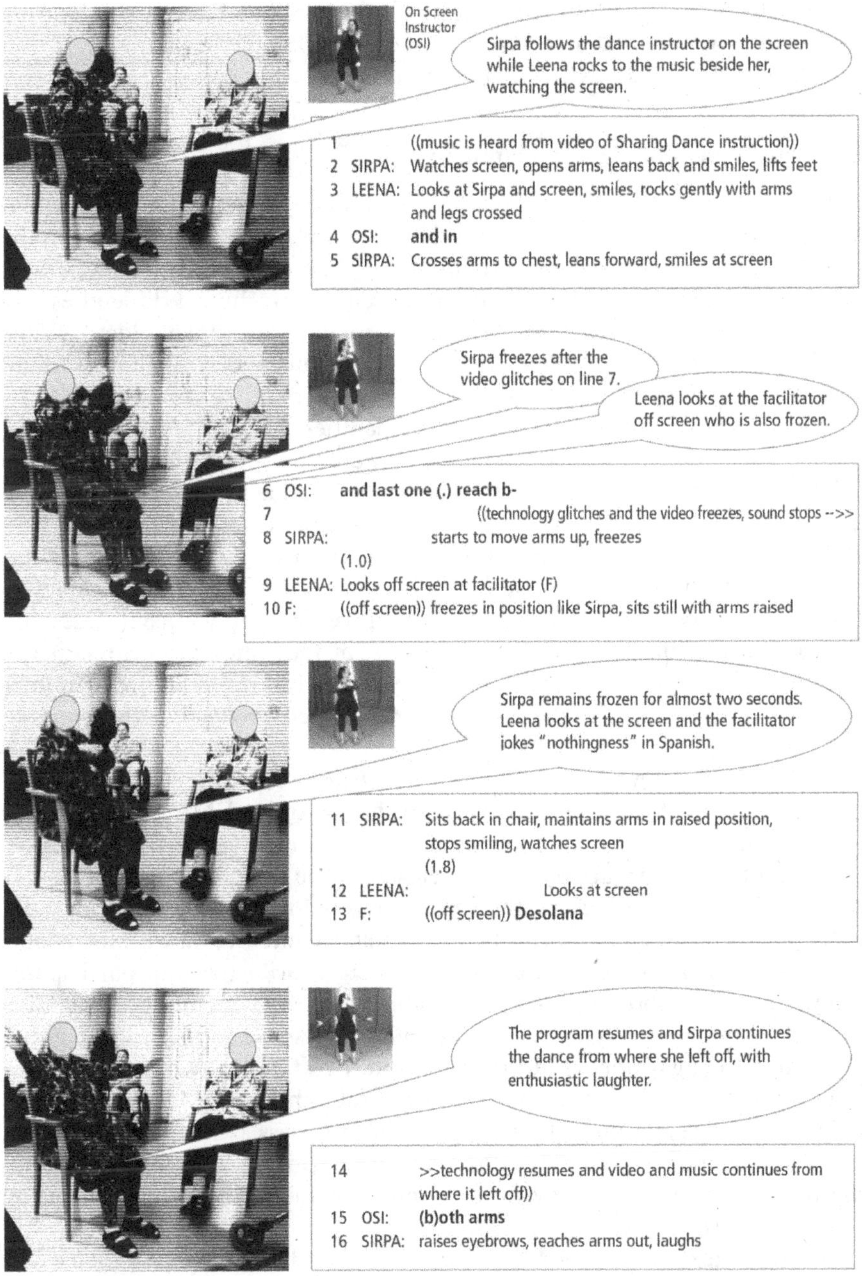

Figure 8.3 Example of EMCA transcript analysis for an international pilot study in Finland.

Sirpa remains focussed on the collective action of the dance. The Sharing Dance program enables a connectedness to the OSI for Sirpa and may also provide relational resources for other co-present participants.

Discussion and concluding comments

In this chapter, we examined the role of technology and discussed its potential limitations in terms of the glitch that interrupts the engagement with the online dance program. What participants said about their experiences drew our attention, in this chapter, to the prevalence of technical difficulties and we found that these mattered differently to different stakeholders. While older adult participants felt these were minor disturbances, facilitators found the technological aspects of the program to be a source of distress.

From our observations of technical difficulties during dance sessions, we found that glitches provided the impetus for creative interactions and opportunities for participants to connect in the room. Facilitators responded in their roles to continue to lead classes from memory and by improvising in creative ways to keep the Sharing Dance program going beyond the limitations of the technology. This exemplifies the effectiveness of NBS' strategy to make dance accessible to people with cognitive and physical challenges and in broader communities. Continued support of facilitators and carers in the technological aspects of the program along with training opportunities that support creativity in leading activities will be beneficial as the program expands further.

In introducing a new branch of study that seeks to explore internationalization of the program, we demonstrated the use of an EMCA analytic lens to also look at how the glitch was made relevant and creatively dealt with by participants in the group. By looking closely at the interaction of one participant with the technology of the screen, we could see that the engagement with the on-screen program was prioritized above interacting with the people in the room. This raises questions of how focus is prioritized for participants in hybrid digital and co-present interactions and how facilitators approach the task of encouraging interaction among co-present members of the group in parallel to engagement with the program. In situations where technology is a part of the environment supporting interactions, the EMCA approach is helpful for understanding how social connectedness is maintained among participants in complex circumstances with multiple foci, as well as informing the design of technology in the context of older adult social inclusion (Arminen, 2017).

In the context of other studies, our results build on existing evidence that effective delivery of ICT for the delivery of programs to support social inclusion will be dependent upon the digital infrastructure and systems in place to support its uptake including training of staff and carers (Van der Heide et al., 2012). Limitations of our study in terms of technology were in the data collected that pertained more to delivery of the program and not explicitly about the use of technology itself. We analysed the data provided in what people said, the words they chose, and the words used in field notes during the data collection. Field notes were limited to when the program began and ended and observations were not recorded of facilitators setting up the program, for example. Much of our data on how the delivery affected the program was limited to interviews with participants and facilitators and relied on how they articulated their responses. Many participants would focus on the OSI when responding to questions about "the delivery of the program." The technology was a minor focus in the evaluation objectives of the program for this study; however, it emerged as an important topic of interest during the research. In future studies, purposefully observing processes where facilitators interact with technology such as setting up equipment and logging on to systems or dealing with technical difficulties may inform the implementation of remote programs with sensitivity to the contexts of different settings. In order for programs like Sharing Dance to enhance a meaningful social life of older adults including people living with dementia and in rural areas, considerations need to be made for the local context of people in places where they intend to engage.

References

Arminen, I. (2017). *Institutional Interaction: Studies of Talk at Work*. Abingdon, UK: Routledge.

Arminen, I., and Poikus, P. (2009). Diagnostic reasoning in the use of travel management system. *Computer Supported Cooperative Work*, *18*, 251–276. https://doi.org/10.1007/s10606-008-9086-3

Arminen, I., Licoppe, C., and Spagnolli, A. (2016). Respecifying mediated interaction. *Research on Language and Social Interaction*, *49*(4), 290–309. https://doi.org/10.1080/08351813.2016.1234614

Braun, V., and Clarke, V. (2006). Using thematic analysis in psychology. *Qualitative Research in Psychology*, *3*(2), 77–101. https://doi.org/10.1191/1478088706qp063oa

Chen, Y. R. R., and Schulz, P. J. (2016). The effect of information communication technology interventions on reducing social isolation in the elderly: A systematic review. *Journal of Medical Internet Research*, *18*(1), e18. https://doi.org/10.2196/jmir.4596

Heritage, J. (1984). *Garfinkel and Ethnomethodology*. Cambridge, UK: Polity Press.

Kamunen, A. (2019). How to disengage: Suspension, body torque, and repair. *Research on Language and Social Interaction*, *52*(4), 406–426. https://doi.org/10.1080/08351813.2019.1657287

Kilpeläinen, A., and Seppänen, M. (2014). Information technology and everyday life in ageing rural villages. *Journal of Rural studies*, *33*, 1–8.

Kosurko, A., Herron, R. V., Skinner, M. W., Bar, R. J., Kontos, P., Grigorovich, A., and Menec, V. (2020a). Opportunities and challenges of digital delivery of sharing dance seniors for social inclusion. *Romanian Journal of Communication and Public Relations*, *22*(2), 23–37. https://doi.org/10.21018/rjcpr.2020.2.298

Kosurko, A., Skinner, M., Herron, H., Bar, R., Grigorovich, A., Kontos, P., and Menec, V. (2020b). Rural gerontechnology: Arts-based insights into rural ageing and the use of technology. In M. Skinner, R. Winterton, and K. Walsh (Eds.), *Rural Gerontology: Towards Critical Perspectives on Rural Ageing*. London: Routledge.

Kosurko, A., Arminen, I., Herron, R., Skinner, M., and Stevanovic, M. (2021). Observing social connectedness in a digital dance program for older adults: An EMCA approach. *International Conference on Human-Computer Interaction* (pp. 393–404). Cham: Springer. https://doi.org/10.1007/978-3-030-78108-8_29

Kosurko, A., Herron, R. V., Grigorovich, A., Bar, R. J., Kontos, P., Menec, V., and Skinner, M. W. (2022). Dance wherever you are: The evolution of multi-modal delivery for social inclusion of rural older adults. *Innovation in Aging*, *6*(2). https://doi.org/10.1093/geroni/igab058

Lerner, G. H., and Raymond, G. (2021). Body trouble: Some sources of difficulty in the progressive realization of manual action. *Research on Language and Social Interaction*, *54*(3), 277–298. https://doi.org/10.1080/08351813.2021.1936994

Moore, R. J. (2013). Ethnomethodology and conversation analysis: Empirical approaches to the study of digital technology in action. In S. Price, C. Jewitt, and B. Brown (Eds.), *The SAGE Handbook of Digital Technology Research* (pp. 217–235). SAGE. https://dx.doi.org/10.4135/9781446282229.n16

O'Connell, M. E., Scerbe, A., Wiley, K., Gould, B., Carter, J., Bourassa, C., and Warry, W. (2018). Anticipated needs and worries about maintaining independence of rural/remote older adults: Opportunities for technology development in the context of the double digital divide. *Gerontechnology*, *17*(3), 126–138. https://doi.org/10.4017/gt.2018.17.3.001.00

Pekkarinen, S., Melkas, H., Kuosmanen, P., Karisto, A., and Valve, R. (2013). Towards a more social orientation in gerontechnology: Case study of the "reminiscence stick". *Journal of Technology in Human Services*, *31*(4), 337–354. https://doi.org/10.1080/15228835.2013.856277

Pinto-Bruno, Á. C., García-Casal, J. A., Csipke, E., Jenaro-Río, C., and Franco-Martín, M. (2017). ICT-based applications to improve social health and social participation in older adults with dementia. A systematic literature review. *Aging & Mental Health*, *21*(1), 58–65. https://doi.org/10.1080/13607863.2016.1262818

Rintel, S. (2013). Video calling in long-distance relationships: The opportunistic use of audio/video distortions as a relational resource. *The Electronic Journal of Communication/La Revue Electronic de Communication (EJC/REC)*, *23*(2). https://www.microsoft.com/en-us/research/wp-content/uploads/2016/02/rintel2013ejcarticle.pdf

Salemink, K., Strijker, D., and Bosworth, G. (2017). Rural development in the digital age: A systematic literature review on unequal ICT availability, adoption, and use in rural areas. *Journal of Rural Studies*, *54*, 360–371. https://doi.org/10.1016/j.jrurstud.2015.09.001

Schegloff, E. A. (2007). *Sequence Organization in Interaction: A Primer in Conversation Analysis (Vol. 1)*. Cambridge University Press. https://doi.org/10.1017/CBO9780511791208

Skinner, M. W., Herron, R. V., Bar, R. J., Kontos, P., and Menec, V. (2018). Improving social inclusion for people with dementia and carers through sharing dance: A qualitative sequential continuum of care pilot study protocol. *BMJ Open*, *8*(11), e026912. https://doi.org/10.1136/bmjopen-2018-026912

Suchman, L. A. (1987). *Plans and Situated Actions: The Problem of Human-Machine Communication*. Cambridge, UK: Cambridge University Press.

Van der Heide, L. A., Willems, C. G., Spreeuwenberg, M. D., Rietman, J., and de Witte, L. P. (2012). Implementation of CareTV in care for the elderly: The effects on feelings of loneliness and safety and future challenges. *Technology and Disability*, *24*(4), 283–291. https://doi.org/10.3233/tad-120359

Walsh, K., O'Shea, E., and Scharf, T. (2020). Rural old-age social exclusion: A conceptual framework on mediators of exclusion across the lifecourse. *Ageing & Society*, *40*(11), 2311–2337. https://doi.org/10.1017/s0144686x19000606

Warburton, J., Cowan, S., and Bathgate, T. (2013). Building social capital among rural, older Australians through information and communication technologies: A review article. *Australasian Journal on Ageing*, *32*(1), 8–14. https://doi.org/10.1111/j.1741-6612.2012.00634.x

Warrener, S., Wrathall, M., Skinner, M., Herron, R., Kontos, P., Menec, V., and Bar, R. (2017). *Improving social inclusion for Canadians with dementia and carers through Sharing Dance: Peterborough pilot phase two report*. Brandon University, Brandon, MB. https://static1.squarespace.com/static/5cc2153bba02c40001991400/t/5e29b489a9d3422148699d9a/1579791499522/SD+P2+Report+2018.pdf

9 On the community dimensions and dynamics of *Sharing Dance*

Verena Menec, Mark Skinner, and An Kosurko

Introduction

Bringing dance to a wide range of communities as a way to foster social inclusion is at the heart of the innovative Sharing Dance Older Adults program offered by Canada's National Ballet School (NBS). As introduced in Chapter 2 by Rachel Bar and John Dalrymple, the evolution of Sharing Dance includes efforts to make the program available to individuals who would typically not have been able to participate in dance, such as individuals living with dementia and their carers, designing content that encourages social engagement among participants, and offering it in a wide range of non-metropolitan regions including rural communities and across the continuum of care settings, both community and institutional. An important aspect of the expansion and sustainability of the program is to build capacity in communities by engaging community organizations and providing dance training and resources, in order to promote social engagement among participants. Accomplishing these goals requires an understanding of *community* and this chapter contributes to the book by focusing on the rural community contexts of Sharing Dance.

Communities are complex. They are the places and spaces in which and through which people create opportunities for shared identity and purpose, sense of belonging, and social participation. Communities differ on a wide range of dimensions. When placed on a rural-urban continuum, for instance, differences in population size and population density matter, which in turn tie into a community's ability to provide services and amenities. Yet, there is also diversity among and within communities. For example, rural regional service centres, that is, larger settlements located in rural regions, can be resource-rich and serve as a resource centre for neighbouring rural areas. In contrast, a small rural village that is located far from a larger service centre may require residents to travel long distances to access services and amenities. And, within every community, there are a diversity of lived experiences and expectations among individuals, households, and groups that contribute to the complex dynamics through which community identity, belonging, and participation occur. The latter raises the importance of social inclusion, which, as discussed in Chapter 5 by Rachel Herron, Sheila Novek, and Verena Menec, has been

DOI: 10.4324/9781003260691-9

conceptualized in many different ways (including in relation to participation, isolation, and loneliness) but most helpfully as a multi-dimensional process that takes into consideration the broader community environment or societal factors that contribute to the social inclusion (or exclusion) of individuals or groups (Walsh et al. 2019).

Consideration of these differing community dimensions and dynamics was central to the *Improving Social Inclusion for People with Dementia and Carers through Sharing Dance* project that is the focus of this book, particularly in regards to the collaborative objective to identify the challenges of expanding the program (e.g., programmatic and contextual factors that influence the success or failure of the program) in terms of improving social inclusion (see Chapter 3, Skinner and Bar). In contributing to this objective, in this chapter, we aim to explore how the implementation of arts-based social inclusion programs like Sharing Dance are embedded within community dimensions and dynamics, and how lessons learned from the creative collaboration undertaken by the research team can help with the sustainability and scaling up of arts-based programs in diverse settings, including in the rural and small town (or non-metropolitan) contexts in which the Sharing Dance program was piloted. In doing so, we draw on the perspectives of staff and volunteers from community organizations involved in the Sharing Dance research collaboration to build knowledge about the implementation of arts-based programs for social inclusion in rural contexts.

The rural context

The discourse on rural areas is often polarized (Keating, 2008). At the one extreme, rural communities – sometimes defined as those areas left over when urban centres have been accounted for (Dijkstra and Poelman, 2014) – are described from a decline and deficit perspective, particularly as they are compared to urban settings. At the opposite end of the spectrum, rural life is sometimes romanticized and portrayed as idyllic and pastoral (Shucksmith, 2018). The growing scholarship on rural ageing suggests a much more complex, dynamic, and nuanced reality (Skinner et al., 2021). The changing demographic, social, and economic contexts of rural communities present both challenges and opportunities for older adults that are different from those in urban settings.

Rural populations have declined worldwide as a result of urbanization and out-migration of young individuals (Berry, 2021). Coupled with older adults' preference to age-in-place, this has led to many (albeit not all) rural communities and regions to experience more rapid increases in older populations than urban centres (Dandy and Bollman, 2008). Social connections can be strong in rural areas (Menec et al., 2015), and there may be a greater attachment to place, and sense of connection to the community than in urban centres (Heley and Woods, 2021). Declining populations, limited economic resources, and large distances, however, have resulted in human and infrastructure deficits in many rural areas (Hanlon et al., 2007). For example, attracting and retaining health human resources, including physicians is an on-going challenge in some rural communities (Fleming and Sinnot, 2018). Yet, research indicates that this is not the case in all rural areas (Menec et al., 2010), suggesting that access to services depends on the specific context and service delivery model. Similarly, some studies show that rural residents are more satisfied with their lives than their urban counterparts (St John et al., 2021), but that there is also considerable variation within rural, as well as urban regions (Helliwell et al., 2019). Inconsistencies in findings regarding rural and urban differences have also emerged for health-related measures, such as functional status and disability (Zhao et al., 2019). Overall, research highlights that both rural and urban areas are diverse and generalizations about rural residents' quality of life are risky.

Large distances, or what has rather vividly been described as the "tyranny of distances" with reference to Australia (Blainey, 2001), create mobility problems. Transportation options for older adults who do not drive may therefore be limited and costly (Novek and Menec, 2014), which can create difficulties in accessing services (Ryser and Halseth, 2012). Moreover, the digital divide is a reality in many rural areas, where internet access is sometimes not available or at best slow and expensive (Haight et al., 2014); even cell phone connectivity may be spotty. This, too, is not universally the case, however, as some rural regions have been able to develop their own high-speed internet connectivity that is on par with that in urban settings.

Not only are rural communities diverse, but they are dynamic: populations can grow or decline; become older as younger individuals leave or older adults move in, such as when a community becomes a retirement community; economic circumstances can improve or decline, for example, due to broader economic booms or bust, or when new industries open up or existing ones close, and so forth. These changes can lead to divergent visions of what the community should look like (Skinner and Winterton, 2018); for example, a small town that increasingly becomes a retirement community can provide more opportunities for healthy, active older adults, but may also marginalize those who are less healthy who can feel increasingly excluded (Keating et al., 2013).

The arts in rural contexts

The arts range the gamut from performing arts, such as dance, music, and theatre, to visual arts, such as painting, to literary arts, such as creative writing (Anwar-McHenry, 2011a). All these arts are as much a presence in rural settings as urban ones. However, their purpose and characteristics can differ because they are performed in a rural context (Duxbury and Campbell, 2011). The literature on the arts in rural areas often distinguishes between three types of art activities: art that originates within communities themselves, festivals, and touring activities. The reality is much more complex than this simple classification would suggest; however, for example, arts programs may be a combination of locally based and visiting (Duxbury and

Campbell, 2011). Sharing Dance, which draws on both external (Canada's National Ballet School) and local resources could be considered as falling into this combined category.

Creating a vibrant arts scene has been used as a community development approach in some rural communities, and can play an essential role in staving off economic and population decline and ensuring communities' sustainability by attracting artists, tourists, or retaining youth (Duxbury and Campbell, 2011). Apart from their economic impacts, the literature has focused on the social benefits of the arts. The arts have been described as fostering feelings of belonging to the community, social connections, empowerment, and collaborations (Balfour et al., 2018). However, as is the case for many other programs or services in rural settings, arts activities hinge on volunteering (Skippington 2016). This may create challenges for sustainability if volunteer support fades. Similarly, maintaining funding is key for the sustainability of arts projects (Anwar-McHenry, 2011a), but maybe more difficult for rural communities than urban centres (Gallagher, 2021).

Community perspectives on Sharing Dance

The *Improving Social Inclusion for People with Dementia and Carers through Sharing Dance* project involved qualitative pilot studies in two non-metropolitan regions of Canada: Peterborough, Ontario and Brandon, Manitoba. As outlined by Mark Skinner and Rachel Bar in Chapter 3, the Peterborough Pilot Study and Brandon Pilot Study examined the delivery of Sharing Dance in ten rural communities located throughout the regions. Not including the urban centres involved in the project are the small towns and villages of Apsley, Ennismore, Havelock, Lakefield, and Millbrook in Peterborough, Ontario; and Carberry, Hamiota, Killarney, Minnedosa, and Neepawa in Brandon, Manitoba (see Figure 3.3 and Figure 3.4 in Skinner and Bar, Chapter 3). These regions were selected because their ageing populations, various urban, small town and rural settlements, and continuum of care settings for older adults (Statistics Canada, 2022) are typical of what is considered non-metropolitan Canada where NBS is seeking to expand the reach of its Toronto-based programming. The regions were also selected due to their proximity to researchers from nearby universities (including the co-authors of this chapter) whose network of research partners, such as Alzheimer Society, Community Care, and LTRC organizations, were keen to participate in the Sharing Dance program as illustrated in Figure 9.1 and Figure 9.2 which are visual examples of community engagement.

In this chapter, we focus on the findings from the multi-phased pilot study research design, particularly from interviews with community organization staff and volunteers and from the reflective notes from research team meetings that were held regularly over the course of the project (see Table 3.1 in Skinner and Bar, Chapter 3). We also reviewed research field notes from the in situ perspective of research team members created while carrying out the pilot studies in Peterborough and Brandon. In combination, these data

Figure 9.1 Sharing Dance recruitment poster, Peterborough pilot study.

provided a rich source of information to critically examine the implementation of Sharing Dance in rural contexts from a community perspective. Analytic themes were derived through an iterative process that involved in-depth reading of transcripts, field notes, and minutes, and discussion among the three authors of this chapter. In presenting our findings, we focus here on three broad, inter-related emergent themes: first, *resource gaps versus resilience*; second, *power dynamics and gatekeepers*; and, third, the *role of the researchers* in the implementation of the Sharing Dance program.

Figure 9.2 Sharing Dance recruitment poster, Brandon pilot study.

Resource gaps versus resilience

While many rural communities are faced with resource gaps, many have also learned to deal with and overcome these gaps. It is therefore important to juxtapose the very real challenges that rural communities face with their resilience and resourcefulness in dealing with these challenges. Resilience can be conceptualized as an individual's or a community's ability to adapt to and thrive despite adversity (Wild et al., 2013). Community resilience can be aided by a variety of factors, such as social capital (e.g., networks, mutual exchange) and physical capital (e.g., infrastructure) (Sapountzaki, 2007). In her research on the arts in rural communities, Anwar-McHenry (2011a) argued that arts programs promote communities' resilience by fostering individuals' sense of place, civic participation, and social participation. Similarly, Roberts and Townsend (2016) discuss how the cultural capital derived from the "creative economy" can promote rural communities' resilience.

Resource gaps (human capacity, infrastructure), but also resilience, were evident in the Sharing Dance project. Lack of capacity prevented a community organization from participating, and capacity challenges and competing demands may have contributed, in part, to another community organization's decision to withdraw from the program. Offering the program further added to the workload of staff who felt already stretched. One facilitator noted in an interview:

> The workload was a little unbalanced due to other factors [at the office]. If the program were to move forward, training other facilitators would be beneficial to lightening the workload.

Conversely, organizations with sufficient capacity were in a better position to participate in the project. As was observed in team reflective notes, long-term residential care (LTRC) institutions "seem to be particularly excited by the program" because it "may be easier resource-wise in LTRC than the community to run this program." Although the reference to resources would also include other aspects such as space and internet access, having staff available, including, as many LTRC already do, recreation facilitators whose job it is to organize and provide activities for residents, would certainly help with the ability to deliver the program.

Lack of equipment (e.g., TV) and internet were barriers in several sites, but staff were resourceful in dealing with these challenges. This resilience was reflected in an interview with a staff member who indicated having applied for funding: "[Name] forwarded me the cost, so I did put a community grant in. I don't have the internet cost figured out yet, so…"

Similarly, communities' ability to overcome lack of staff capacity was evident. In this respect, consistent with the rural literature (Menec et al., 2022; Skinner et al., 2014), volunteers played a key role in filling capacity gaps; for example, volunteers helped to facilitate classes. By helping to create social connections among program participants, they therefore contributed to the success of Sharing Dance by fostering social connections among *individuals*. At the same time, their involvement contributed to *community* social inclusion. The example of one site will serve to illustrate this point.

The site, a LTRC institution, is located in a small rural town with a population of fewer than 1,000 residents. The town has a number of amenities, including a health centre with a hospital and emergency department that serves the surrounding area, a community centre, and an art centre with an exhibition gallery. The town has a long-standing focus on older adults, as reflected in the fact that it joined the provincial age-friendly community initiative over a decade ago. This initiative aims to have communities throughout the province work on enhancing the social and physical environment to promote health and quality of life of older adults. As a participant in a regional collaboration that developed its own internet provider capacity, the town enjoys excellent internet access. Thus, despite its small size, the town has many strengths to draw on.

The recreation program coordinator at this site was enthusiastic about the program coming to the institution. She said in an interview:

> I was just so thrilled that you guys chose to come to [name of town]. As soon as I found out about it I thought this is cool. I didn't know what to expect but I was just so happy.

Within the context of an already well-established volunteer culture in the community, as well as a strong focus on older adults, this enthusiasm may have set the stage for even greater volunteer involvement than normal. As the program facilitator explained in response to an interview question of what they liked the most about being a facilitator:

> To have that much volunteer involvement I would say because I'm responsible for 30 residents, so when we have a crew that large and they're all out like that – which sometimes I average about 20 people – sometimes I can't be the motivator for each one in particular when I'm trying to run a full program. So when we have those volunteers serve a line of people who need that extra boost to keep going, that was an asset. I don't always have that many volunteers on hand. Which is great motivation so therefore they're able to do more and be more active.

The volunteer involvement was thought to have contributed to making the program more enjoyable for participants and to have contributed to their well-being, but also benefitted the volunteers and created new social connections for them. This is consistent with research that shows the benefits of volunteering for the volunteers, including the social, mental, and physical health benefits (Gottlieb and Gillespie, 2008; Pettigrew et al., 2019; Smith et al., 2018). As one volunteer noted:

> Well – It was fun for me to go and be with these older people – I'm not from this area, so I don't have relatives – I know [name] and a few from around town. It was a place I would never go, so I was glad to have that chance to be part of it.

When further asked about whether the program has been a good way to connect with other people, the volunteer responded:

> Yes, because the volunteers were ever so happy – they enjoyed it immensely – they got physical movement, good for them, the volunteers, the residents just enjoyed having all these volunteers around dancing away and it was a very positive experience for everybody I think. I think they enjoyed having the volunteers. It's not often you get a lot of people come and doing something like that. And laughing you know.

The community spirit was not restricted to staff and volunteers directly involved in Sharing Dance but extended to other staff as well. One staff member commented in an interview: "I couldn't believe even the kitchen staff were doing it." This sense of community between volunteers, the LTRC institution residents and staff, continued beyond the class itself, as reflected in the following comment from an interview with a facilitator:

> One volunteer is bringing in lemon pies on fair day which she wouldn't have thought of if she hadn't spent 8 weeks with us. She thought she wanted to do something but we couldn't take them all to the fair. It sparked something in the volunteers. I could see bits of that and that was nice too. I think because they spent so much time together.

While the extent of volunteering was greater in this case example than in any other site, the case helps to illustrate several key issues that contributed to the success of the program: 1) having a local person who champions the program helps; 2) having a culture of volunteering – or high social capital – in the community is of benefit, as it provides a basis to draw volunteers into the program; and 3) the program has benefits not just for participants, but also volunteers, and helps to build community social inclusion by creating connections between participants, volunteers, and staff. As such it can help bridge the divide that sometimes exists between community members and residents in LTRC institutions.

Power dynamics and gatekeepers

Power dynamics need to be considered in any collaboration as they can affect the implementation and, ultimately, the success of programs. Power – at its most basic defined as the ability to influence others – is relational and embedded within social interactions and the context in which they occur. A discussion of the concepts of power and empowerment, which have been explored from a wide range of disciplinary perspectives and generated vast bodies of literature (e.g., Anderson and Brion, 2014; Foucault, 1980; Lukes, 2005; Orford, 1992), is beyond the scope of this chapter; here we more pragmatically focus on how power dynamics and gatekeepers – influential individuals or groups that have the power to either restrict or provide access to a resource – manifested themselves in the *Improving Social Inclusion for People with Dementia and Carers through Sharing Dance* research project, and what can be learned from our experiences.

As outlined earlier in this book (see Herron and Skinner, Chapter 1; Skinner and Bar, Chapter 3), the project involved multiple layers of creative collaborations between organizations and individuals working within them: NBS, universities, community organizations, active living centres, and LTRC institutions; individuals involved included professional dance instructors, researchers, administrators in community organizations, staff, and volunteers, who, in turn connected with participants, but also other individuals not directly involved in the Sharing Dance project, such as other staff in organizations. As such, a large web of relationships were in play, all of which were embedded within a broader organizational and rural context.

The rural context of the project may have provided a certain advantage in bringing Sharing Dance into communities, as rural places are often marginalized in terms of arts-based programming (Gallagher, 2021), as well as research. Thus, having NBS, a national, highly recognized organization as well as two universities approach community organizations may have opened doors, as reflected in the following two quotes from interviews with staff:

> I think for a small place, this was really, really special. ... Small towns are most often ...not chosen for things like this.
>
> I like how they reached out to different communities instead of just [name of site]. People were quite excited. The research was exciting.

Designed collaboratively between NBS and the researchers, and implemented in close collaboration with community partners, the research provided a continuous feedback cycle that helped promote an equitable partnership, which benefited the implementation of the Sharing Dance program. For example, early in the research project, recruiting participants for the program was a task taken on by the partnering community organization. This was felt to be onerous, and the suggestion was made that the researchers should take on some of this work in one follow-up interview with staff.

> I would like to see more involvement in program promotion through the university and NBS, as myself, my volunteers who did a lot of outreach to encourage people to come, and if I was to tally up those hours it would probably would be a full week's work or more.

The feedback led the researchers to take a more active role in participant recruitment in the next pilots, for example, by making presentations at gatherings. As one facilitator noted in an interview in a subsequent phase of data collection:

> The open house beforehand was a great idea – there was a few families who did come and got to see, ask questions, [the researcher] explained it, just to meet them. I thought that was great, we didn't have that last time.

Similarly, it became quickly apparent that offering the Sharing Dance program via live streaming was too challenging, given the problems with internet connectivity in rural communities (see Kosurko, Arminen, and Stevanovic, Chapter 8 in this book for a more in-depth discussion). This led to the shift in how the program was offered; a pre-recorded option for download was provided in the short term, and eventually the format shifted to a pre-recorded streaming format. In order to retain the personal connection between the on-screen dance instructors (OSI) and participants, facilitators were encouraged to provide information that would be meaningful to participants, such as mentioning a birthday, which was appreciated (Kosurko et al., 2022). The importance of having a facilitator in the room also became apparent, which led NBS to expand their focus on facilitator training as part of the evaluation of Sharing Dance (for a more detailed discussion, see Bar and Dalrymple, Chapter 2).

The gatekeeper phenomenon is not restricted to rural settings, but the influence of one individual can be greater in rural contexts than in urban centres, given close relationship networks and limited opportunities to bypass individuals in positions of power. One individual therefore has the potential to stall a program from getting off the ground (Menec et al., 2014). Moreover, power dynamics within a community – the "small town politics" – whereby certain persons (or groups) have conflicting ideas or personalities can be detrimental (Menec and Brown, 2018). These dynamics were apparent at one site, where an influential person in the community tried to stall the recruitment of participants. It emerged that the concern was that participating in a program designed for people living with dementia would "out" individuals as having dementia. Thus, the gatekeeping behaviour was the result of the stigma associated with having dementia. Ultimately, the issue was resolved because the research associate was at the site and became aware of the problem and was subsequently able to help recruit participants. This experience also highlights the importance of having a person on-site to help navigate problems as they arise.

Power dynamics also come into play in interpersonal relationships. In one instance, an interpersonal conflict emerged between a program coordinator who was new to the position and a facilitator:

> First day [of the program] I was there, my equipment was in her office, I didn't have the password, she closed her door, shut blinds, the body language whether she meant it, she was very closed. She wasn't a greeter or offering of help.

The conflict made it difficult for the facilitator to hold classes. These interpersonal issues are difficult to foresee and deal with; open communication between the different members of the team may help resolve such issues.

Researcher involvement in the Sharing Dance program

By studying Sharing Dances, the *Improving Social Inclusion for People with Dementia and Carers through Sharing Dance* project was not only instrumental in its development and modification (e.g., the shift from live streaming to pre-recorded streaming), but it became an integral part of its implementation. This means that the implementation of the program cannot be discussed without considering the role that the research team played. While the research created some additional complexities, it also provided support for community organizations that were experiencing capacity limitations.

Researchers' involvement started with engaging the community organizations that could deliver the program. In other words, organizations did not simply come forward on their own to volunteer their participation, but rather researchers had to actively recruit them. This was easier in Peterborough, Ontario than in Brandon, Manitoba, as one of the researchers could draw on an already established relationship with a community organization that offered services in multiple sites. As such, it was easy to launch the program in several sites. In Manitoba, new relationships had to be established, not only between researchers and community organizations but also between community organizations themselves, which was more time-consuming.

Given the nature of the collaborative project – part dance program, part research – advertising and explaining the two parts created some challenges. Moreover, while it is difficult to gauge the impact of adding the research component to the program, it may have prevented some individuals from wanting to participate in the dance program, given the additional tasks, including informed consent forms, interviews, focus groups, diaries, and observations.

However, having a research component was also beneficial as it provided the capacity to deliver the program. The researchers could help problem solve, for example, by helping to find appropriate space for the classes. The research project was also able to fill resource gaps, by providing the funds for TVs and laptops for some sites. Moreover, having the research associate attend classes for the observation component of the project was welcomed, as she could also help deal with issues as they arose, such as computer issues. One facilitator stated in an interview: "I really did feel a lot of support from you [research associate], you're never too sure – I always knew I could count on you."

Discussion and lessons learned

In this chapter, we aimed to describe how community dimensions and dynamics can influence the implementation of arts-based programs and how lessons learned from the Sharing Dance program in rural communities of Ontario and Manitoba can help with sustainability and scaling up such programs. Based on the findings presented above, we offer four inter-related lessons learned for discussion alongside the promising practices identified elsewhere (see Skinner and Bar, Chapter 3) and in the concluding chapter of the book (Bar and Herron, Chapter 10).

Being aware of the local context

Rural communities are diverse. Although rural communities experience many challenges, their resilience to dealing with these challenges should not be overlooked. Local strengths, such as a vibrant volunteer culture, can be a major asset and can be built upon in launching an arts program. Conversely, being cognizant that staff or volunteers are already stretched to the limits suggests the need to provide additional supports, such as relevant training opportunities or helping secure funding. Capacity may also fluctuate over time. For example, being mindful that certain times of the year may be more stressful for community organizations who already work with limited resources, such as the financial year-end, suggests the need to be open to adjust timelines to meet the needs of community partners (see Bar and Kontos, Chapter 4).

Promoting the arts in rural contexts

The arts add to the vibrancy of communities and may help with the sustainability of rural communities (Duxbury and Campbell, 2011). Although the arts are very much a presence in rural communities, they can be undervalued (Skippington, 2016), particularly in comparison with other activities like sports (Anwar-McHenry, 2011b). However, bringing arts programs to rural communities provides the opportunity to attract and "bring out of the woodwork" (Anwar-McHenry, 2011b, p. 42) segments of the population whose interests may not normally be served by community activities. It may also draw arts-minded individuals to settle in a community (Gibson and Gordon, 2018). Organizations like NBS play a key role in bringing arts-based programs to rural communities and promoting the arts in general. In doing so, creative collaboration with community organizations is essential, as the *Improving Social Inclusion for People with Dementia and Carers through Sharing Dance* project showed. Such organizations know the local context, can tap into local strengths, and can anticipate potential challenges, all of which helps with the sustainability of programs.

The potential challenges of promoting an arts-based program like Sharing Dance as art versus exercise needs to be acknowledged at this point. Chapter 4 by Rachel Bar and Pia Kontos in this book addresses this issue in more depth; suffice it to say here that there remains a tension between promoting dance (or indeed other arts, such as music or the visual arts) as a creative art versus therapy to promote health. This tension becomes particularly evident when it involves older adults, and even more so individuals living with dementia, given the prominence of the biomedical model of ageing and the medicalization of ageing. There is no easy solution to resolve this tension. Indeed, applying for funding may require a "reframe" whereby an arts-based program is presented as providing health benefits in order to secure a grant, which was the case in the Public Health Agency of Canada (PHAC) funding that NBS has received in support of the expansion of the program (see Bar and Dalrymple, Chapter 2). However, this does not preclude presenting the program first and foremost as art.

Addressing power inequities between partners

Power dynamics in partnerships are complex and occur not only between organizations but also between individuals; they are embedded within contexts, such as structural factors that marginalize certain groups or communities (Wallerstein et al., 2019). The sustainability and successful scaling up of arts-based community-based programs hinges on establishing equitable partnerships between arts organizations and community partners. Being aware of the local context in which the program is offered is critical, as noted above. Establishing an equitable partnership structure that fosters open communication is also key. This may include providing opportunities for "deliberative communication" on the partnership process to help reduce power differentials (Wallerstein et al., 2019). An open dialogue regarding goals and expectations can further reduce possible misunderstandings and resulting conflicts. Furthermore, supporting community partners with resources, such as facilitator training, can help redress inequities.

Supporting arts-based programs with research

The *Improving Social Inclusion for People with Dementia and Carers through Sharing Dance* project afforded the unique opportunity to not only evaluate the implementation process, and whether expected goals were achieved, but also allowed addressing broader research questions, such as the potential of dance programs to support embodied self-expression, creativity, and social engagement of persons living with dementia (Kontos et al., 2021). Thus, research goes beyond an evaluation of the program per se and advances the theorizing and empirical knowledge base on the arts and related areas, such as the social inclusion of individuals living with dementia. As outlined in the present chapter, conducting research can also provide much-needed on-the-ground support to communities in offering a program. As discussed, in the Sharing Dance project, the research component could not be disentangled from the implementation of the dance program – and we would argue that this should not be seen as problematic; rather, as the literature on the benefits of participatory research shows (Cargo and Mercer, 2008; Wallerstein et al., 2018), research ideally occurs in tandem with program implementation.

Concluding comments

The *Improving Social Inclusion for People with Dementia and Carers through Sharing Dance* project provided several key insights into community dimensions and dynamics that could inform any future considerations regarding the sustainability and scaling up of arts-based programs in rural contexts. The project further contributes to the literature on social inclusion by highlighting the multi-dimensional nature of social inclusion, with the community context, organizational, as well as interpersonal factors impacting the implementation of Sharing Dance and, ultimately, the social inclusion of program participants, their carers, and volunteers. Being mindful of resource gaps in rural communities, as well as building on existing community strengths is, therefore, important. Moreover, power dynamics were evident in our research, suggesting the need to build equitable partnerships between arts organizations and community organizations. Partnership processes need to allow the space for open communication about possible issues and tensions that may arise within and between organizations. While such communication takes time, it is time well spent as it will contribute to the success of the program and promote social inclusion, as well as fostering the creative collaborations such as those at the heart of this chapter and book.

References

Anderson, C., and Brion, S. (2014). Perspectives on power in organizations. *Annual Review of Organizational Psychology and Organizational Behavior*, *1*(1), 67–97. https://doi.org/10.1146/annurev-orgpsych-031413-091259

Anwar-McHenry, J. (2011a). Rural empowerment through the arts: The role of the arts in civic and social participation in the Mid West region of Western Australia. *Journal of Rural Studies*, *27*(3), 245–253. https://doi.org/10.1016/j.jrurstud.2011.03.001

Anwar-McHenry, J. (2011b). “They’d rather go play footy”: An exploratory study of the enjoyment and benefits of the arts in Western Australia’s remote Murchison region. *Geographical Research*, *49*(1), 37–46. https://doi.org/10.1111/j.1745-5871.2010.00646.x

Balfour, B., Fortunato, W.-P. M., and Alter, T. R. (2018). The creative fire: An interactional framework for rural arts-based development. *Journal of Rural Studies*, *63*, 229–239. https://doi.org/10.1016/j.jrurstud.2016.11.002

Berry, E. H. (2021). Demographic ageing and rural population change. In M. Skinner, R. Winterton, and K. Walsh (Eds.), *Rural Gerontology: Towards Critical Perspectives on Rural Ageing* (pp. 17–28). Routledge: Abingdon, UK.

Blainey, G. (2001). *The Tyranny of Distance: How Distance Shaped Australia’s History*. Sydney: Macmillan.

Cargo, M., and Mercer, S. L. (2008). The value and challenges of participatory research: Strengthening its practice. *Annual Review of Public Health*, *29*(1), 325–350. https://doi.org/10.1146/annurev.publhealth.29.091307.083824

Dandy, K., and Bollman, R. D. (2008). Seniors in rural Canada. *Rural and Small Town Canada Analysis Bulletin*, Catalogue No. 21-006-X. Ottawa: Statistics Canada.

Dijkstra, L., and Poelman, H. (2014). *A harmonized definition of cities and rural areas: The new degree of urbanisation*. Regional Working Paper, 01/14, Directorate-General for Regional and Urban Policy, European Commission.

Duxbury, N., and Campbell, H. (2011). Developing and revitalizing rural communities through arts and creativity. *Small Cities Imprint*, *3*(1), 111–122.

Fleming, P., and Sinnot, M. L. (2018). Rural physician supply and retention: Factors in the Canadian context. *Canadian Journal of Rural Medicine*, *23*(1), 15–20. https://doi.org/10.1111/j.1748-0361.1992.tb00351.x

Foucault, M. (1980). *Power/Knowledge: Selected Interviews and Other Writings 1972-1977*. C. Gordon (Ed.). New York: Pantheon Books.

Gallagher, B. K. (2021). Death and life: The promise and problems of the arts in rural communities. *International Journal of Arts Management*, *23*(2), 30–42.

Gibson, C., and Gordon, A. (2018). Rural cultural resourcefulness: How community music enterprises sustain cultural vitality. *Journal of Rural Studies*, *63*, 259–270. https://doi.org/10.1016/j.jrurstud.2016.11.001

Gottlieb, B. H., and Gillespie, A. A. (2008). Volunteerism, health, and civic engagement among older adults. *Canadian Journal on Aging*, *27*(4), 399–406. https://doi.org/10.3138/cja.27.4.399

Haight, M., Quan-Haase, A., and Corbett, B. A. (2014). Revisiting the digital divide in Canada: The impact of demographic factors on access to the internet, level of online activity, and social networking site usage. *Information, Communication & Society*, *17*(4), 503–519. https://doi.org/10.1080/1369118X.2014.891633

Hanlon, N., Halseth, G., Clasby, R., and Pow, V. (2007). The place embeddedness of social care: Restructuring work and welfare in Mackenzie, BC. *Health & Place*, *13*(2), 466–481. https://doi.org/10.1016/j.healthplace.2006.05.006

Heley, J., and Woods, M. (2021). Rural studies of ageing. In M. Skinner, R. Winterton, and K. Walsh (Eds.), *Rural Gerontology: Towards Critical Perspectives on Rural Ageing* (pp. 29–39). Abingdon, UK: Routledge.

Helliwell, J. F., Shiplett, H., and Barrington-Leigh, C. P. (2019) How happy are your neighbours? Variation in life satisfaction among 1200 Canadian neighbourhoods and communities. *PLoS ONE*, *14*(1), e0210091. https://doi.org/10.1371/journal.pone.0210091

Keating, N. (2008). *Rural Ageing: A Good Place to Grow Old*. Bristol: Policy Press.

Keating, N., Eales, J., and Phillips, J. E. (2013). Age-friendly rural communities: Conceptualizing 'best-fit'. *Canadian Journal on Aging*, *32*(4), 319–332. https://doi.org/10.1017/s0714980813000408

Kontos, P., Grigorovich, A., Kosurko, A., Bar, R. J., Herron, R. V., Menec, V. H., and Skinner, M. W. (2021). Dancing with dementia: Exploring the embodied dimensions of creativity and social engagement. *The Gerontologist*, *61*(5), 714–723. https://doi.org/10.1093/geront/gnaa129

Kosurko, A., Herron, R. V., Grigorovich, A., Bar, R. J., Kontos, P., Menec, V., and Skinner, M. W. (2022). Dance wherever you are: The evolution of multi-modal delivery for social inclusion of rural older adults. *Innovation in Aging*, 6(2). https://doi.org/10.1093/geroni/igab058

Lukes, S. (2005). *Power: A Radical View* (2nd ed). New York: Palgrave Macmillan.

Menec, V. H., and Brown, C. (2018) Facilitators and barriers to becoming age-friendly: A review. *Journal of Aging and Social Policy*, *34*(2), 175–197. https://doi.org/10.1080/08959420.2018.1528116

Menec, V., Nowicki, S., and Kalischuk, A. (2010). Transfers to acute care hospitals at the end of life: Do rural/remote regions differ from urban regions? *Rural and Remote Health*, *10*(1), 1281–1281. https://doi.org/10.22605/rrh1281

Menec, V. H., Novek, S., Veselyuk, D., and McArthur, J. (2014). Lessons learned from a Canadian, province-wide age-friendly initiative: The Age-Friendly Manitoba Initiative. *Journal of Aging & Social Policy*, 26, 33–51. https://doi.org/10.1080/08959420.2014.854606

Menec, V. H., Bell, S., Novek, S., Minnigaleeva, G., Morales, E., Ouma, T., Parodi, J. F., and Winterton, R. (2015). Making rural and remote communities more age-friendly: Experts' perspectives of issues, challenges and priorities. *Journal of Aging & Social Policy*, *27*(2), 173–191. https://doi.org/10.1080/08959420.2014.995044

Menec, V., Newall, N. E. G., Milgrom, R., and Camps, D. (2022). Exploring the sustainability of age-friendly initiatives in a Canadian province. *The Gerontologist*, *62*(1), 18–28. https://doi.org/10.1093/geront/gnab080

Novek, S. and Menec, V. H. (2014). Older adults' perceptions of age-friendly communities in Canada: A photovoice study. *Ageing and Society*, *34*(6), 1052–1072. https://doi.org/10.1017/S0144686X1200150X

Orford, J. (1992). *Community Psychology: Theory and Practice*. Chichester: John Wiley & Sons.

Pettigrew, S., Jongenelis, M. I., Jackson, B., Warburton, J., and Newton, R. U. (2019). A randomized controlled trial and pragmatic analysis of the effects of volunteering on the health and well-being of older people. *Aging Clinical and Experimental Research*, *32*(4), 711–721. https://doi.org/10.1007/s40520-019-01241-3

Roberts, E., and Townsend, L. (2016). The contribution of the creative economy to the resilience of rural communities: Exploring cultural and digital capital. *Sociologia Ruralis*, *56*(2), 197–219. https://doi.org/10.1111/soru.12075

Ryser, L., and Halseth, G. (2012). Resolving mobility constraints impeding rural seniors' access to regionalized services. *Journal of Aging & Social Policy*, *24*(3), 328–344. https://doi.org/10.1080/08959420.2012.683329

Sapountzaki, K. (2007). Social resilience to environmental risks: A mechanism of vulnerability transfer? *Management of Environmental Quality*, *18*(3), 274–297. https://doi.org/10.1108/14777830710731743

Shucksmith, M. (2018). Re-imagining the rural: From rural idyll to Good Countryside. *Journal of Rural Studies*, *59*, 163–172. https://doi.org/10.1016/j.jrurstud.2016.07.019

Skinner, M. W., and Winterton, R. (2018). Interrogating the contested spaces of rural aging: Implications for research, policy, and practice. *The Gerontologist*, *58*(1), 15–25. https://doi.org/10.1093/geront/gnx094

Skinner, M. W., Joseph, A. E., Hanlon, N., Halseth, G., and Ryser, L. (2014). Growing old in resource communities: Exploring the links among voluntarism, aging, and community development: Growing old in resource communities. *The Canadian Geographer*, *58*(4), 418–428. https://doi.org/10.1111/cag.12087

Skinner, M. W., Winterton, R., and Walsh, K. (Eds.). (2021). *Rural Gerontology: Towards Critical Perspectives on Rural Ageing*. Abingdon, UK: Routledge.

Skippington, P. (2016). *Harnessing the Bohemian: Artists as Innovation Partners in Rural & Remote Communities*. Canberra: Australian Nation University Press.

Smith, R., Drennan, V., Mackenzie, A., and Greenwood, N. (2018). Volunteer peer support and befriending for carers of people living with dementia: An exploration of volunteers' experiences. *Health & Social Care in the Community*, *26*(2), 158–166. https://doi.org/10.1111/hsc.12477

Statistics Canada. 2022. (table). Census Profile. 2021 Census of Population. Statistics Canada Catalogue No. 98-316-X2021001. Ottawa. Released April 27, 2022. https://www12.statcan.gc.ca/census-recensement/2021/dp-pd/prof/index.cfm?Lang=E (accessed May 25, 2022).

St John, P., Menec, V., Tate, R., Newall, N., Cloutier, D., and O'Connell, M. (2021). Life satisfaction in adults in rural and urban regions of Canada – The Canadian Longitudinal Study on Aging. *Rural and Remote Health*, *21*(3), 6631–6631. https://doi.org/10.22605/RRH6631

Wallerstein, N., Duran, B., Oetzel, J. G., and Minkler, M. (Eds.). (2018). *Community-Based Participatory Research for Health: Advancing Social and Health Equity* (3rd ed.). San Francisco: Wiley.

Wallerstein, N., Muhammad, M., Sanchez-Youngman, S., Rodriguez Espinosa, P., Avila, M., Baker, E. A., … and Duran, B. (2019). Power dynamics in community-based participatory research: A multiple–case study analysis of partnering contexts, histories, and practices. *Health Education & Behavior*, *46*, 19S–32S. https://doi.org/10.1177/1090198119852998

Walsh, K., O'Shea, E., and Scharf, T. (2019). Rural old-age social exclusion: A conceptual framework on mediators of exclusion across the lifecourse. *Ageing and Society*, *40*(11), 2311–2337. https://doi.org/10.1017/s0144686x19000606

Wild, K., Wiles, J. L., and Allen, R.E. S. (2013). Resilience: Thoughts on the value of the concept for critical gerontology. *Ageing and Society*, *33*(1), 137–158. https://doi.org/10.1017/S0144686X11001073

Zhao, G., Okoro, C. A., Hsia, J., Garvin, W. S., and Town, M. (2019). Prevalence of disability and disability types by urban–rural county classification – U.S., 2016. *American Journal of Preventive Medicine*, *57*(6), 749–756. https://doi.org/10.1016/j.amepre.2019.07.022

10 Future directions for collaborative arts-based ageing research

Rachel Bar and Rachel Herron

Introduction

The aim of this book is to advance collaborative arts-based ageing research through sharing theoretical, methodological, and applied lessons from the four-year nationally funded research project *Improving Social Inclusion for People with Dementia and Carers through Sharing Dance*. The objectives of the project were to understand the experiences, effectiveness, and expansion of Canada's National Ballet School's innovative Sharing Dance Older Adults program, particularly in relation to improving social inclusion of older people living with dementia and carers. The collaborative aspiration of the project brought together researchers, artists, older adults, and community and organizational leaders to offer a more comprehensive view of the impacts of dance in relation to ageing, health, and care. We hope that the overarching lesson from this book is that collaborative arts-based research and programming can have transformative impacts on the lives of older adults, carers, volunteers, and artists as well as communities and institutions. Understanding these impacts and the processes through which they can be achieved is critical to developing sustainable programming, practices, and policies that support the inclusion and well-being of older adults.

In this final chapter, we review the theoretical, methodological, and applied lessons learned from researching Sharing Dance. In the first section, we outline the lessons for researchers in gerontology as well as the impacts of Sharing Dance in relation to challenging stigma and promoting creative opportunities. In the second section, we discuss the next steps in terms of future programming, future research, and putting knowledge into action. We discuss these opportunities by critically examining some of the challenges to sustaining and adapting lessons learned in a world that continues to perpetuate discriminatory attitudes and unequal opportunities in relation to older adults, particularly those living with dementia (Herron et al., 2021; Kontos et al., 2020a).

Theoretical contributions

In the past decade, a small and growing body of research has begun to shed light on the potential of arts-based programs to transform social experiences

DOI: 10.4324/9781003260691-10

and attitudes as well as contribute to broader understandings of health and well-being in later life (Dupuis et al., 2016; Kontos and Grigorovich, 2018; Kontos et al., 2021; Moody and Phinney, 2012). Yet, the bulk of research on ageing, arts, and health has focused on health promotion, disease prevention, and the treatment and management of chronic conditions such as Parkinson's disease and dementia (Fancourt and Finn, 2019). In much of this research, the contributions of the arts are assessed quantitatively; the arts are viewed as an instrument to achieve specific health outcomes; and older adults are viewed as passive recipients of care (Parsons et al., 2017; Zeilig et al., 2019). The experiential impacts of the arts, the processes through which the arts have their effects, and the agency of older adults engaged in the arts have received less attention from researchers. Throughout the chapters of this book, the contributing authors build on these theoretical, empirical, and methodological observations as a point of departure for exploring the potential of Sharing Dance.

Building on the growing recognition that arts-based programs have more holistic health and social impacts, one of the important theoretical contributions of this book is to explore these impacts on multiple levels through the concept of social inclusion. Rather than view social inclusion as just an outcome that is achieved through the reduction of isolation and loneliness and an increase in social engagement at the individual level, the contributing authors built on several recent multi-dimensional frameworks of social inclusion and exclusion (Pinkert et al., 2019; Walsh et al., 2019). These frameworks highlight the resources and systems that shape individual experiences in ways that either promote the rights of older adults to participate in the social world or create barriers to their inclusion (e.g., community dynamics, services and amenities, institutional practices, financial resources, ageism, and stigma). In the second half of this book, the contributing authors explored individual situations and broader systems influencing social inclusion including the embodied experiences of older adults living with dementia, the feelings and meanings that participants associated with Sharing Dance, the importance of community resources in supporting the success of the program, the role and limitations of technology in addressing accessibility issues, and the impact of the program on institutions and social attitudes more broadly.

A key contribution of this book is advancing multi-dimensional frameworks of social inclusion by identifying component parts and how they work together in the context of arts-based programs. In Chapter 5, Rachel Herron, Sheila Novek, and Verena Menec outlined a multi-dimensional approach to social inclusion that centres individual experiences of meaningful participation, situates these experiences within particular places, recognizes that social inclusion is both an outcome and a process, and connects these processes to resources or strategies at multiple levels. The authors illustrated how multi-level processes (i.e., individual, interpersonal, programmatic, community, and institutional) interacted in ways that promoted experiences of social inclusion in Sharing Dance. For example, programmatic emphasis on creativity, exploration, and self-expression over time fostered increased comfort and confidence at the individual level that enabled participants to take risks and be more playful. This, in turn, influenced how family, staff, and volunteers understood diverse older adults at an interpersonal level. These programmatic, individual, and interpersonal processes are difficult to disentangle, and they reveal the complex integration of multi-level processes involved in supporting social inclusion. Continuing to examine the integration of processes at multiple levels in support of social inclusion is a critical direction for future research in this area. By drawing links between individual outcomes and broader processes, researchers can develop the evidence needed to advocate for relevant resources and strategies to promote social inclusion at all levels.

Another important contribution to conceptualizations of social inclusion within this book is the broad range of perspectives and experiences that are included. Social inclusion involves understanding individual experiences of meaningful participation, which are diverse and complex. In Chapter 6, Craig Wingrove, Ruth Snider, Whitney Strachan, and Margo Dunphy shared the feelings, meanings, and processes they associated with Sharing Dance. They explained that part of what made Sharing Dance meaningful and effective was that it offered similar opportunities for joy, social connection, growth, and learning to everyone. They showed that social inclusion is an experiential process that all those involved in the program shared. Some previous research, particularly from an interventionist paradigm, treats social inclusion as something that is done to older adults through opportunities for social engagement (Camic et al., 2014; Dow et al., 2011); however, this chapter showed how the experiences of older adults, artists, and community facilitators are mutually supportive of social inclusion, enriching relational understandings of these processes.

Building on relational and multi-dimensional approaches to social inclusion, Pia Kontos and Alisa Grigorovich, in Chapter 7 linked the individual embodied and creative self-expression of people living with dementia in Sharing Dance to the need for broader institutional and societal change. Their analysis built on developments in critical gerontology related to embodiment and dementia, which emphasize the critical role of the body in selfhood (Kontos, 2012). Rather than privileging what people say and think as central to being a person, this body of research emphasizes the capacity of the body to perceive and engage with the world. At a micro-level, such embodied expression is fundamental to understanding inclusion and well-being, particularly as it relates to people living with dementia. Yet, most social inclusion research privileges what people say and think about inclusion. By focusing on embodied expression in a dance class, this chapter highlights the importance of seeing and supporting the abilities and creativity of people living with dementia. Through creative embodied expression, people living with dementia express themselves, relate to others, and exercise their abilities in ways that others see and value. Based on these observations, Kontos and Grigorovich argue that institutional practices and policies must support access to creative opportunities that promote embodied self-expression and recognize the rights of older adults living with dementia to participate in the social world.

The chapters of the book also investigated meso- and macro-level influences on social inclusion, including the role of technology in addressing social exclusion by facilitating social relations and access to services and programs. Previous research in the field of gerontechnology has begun to examine the role of technology in addressing isolation and exclusion (Chen and Schulz, 2016); however, challenges have been identified in relation to individuals, communities, and organizations that lack technological literacy or infrastructure (Warburton et al., 2013). In Chapter 8, An Kosurko, Ilkka Arminen, and Melisa Stevanovic build on this work by examining the limitations and opportunities associated with the remote delivery of Sharing Dance. In particular, they highlighted the pressure that technological glitches can put on local facilitators in a remotely delivered program, and the need for continued support and training to ensure long-term success of delivering the program remotely. Similarly, Verena Menec, Mark Skinner, and An Kosurko in Chapter 9 pointed to the importance of community resources such as infrastructure, equipment, and volunteers as well as local power dynamics in supporting social inclusion. Both Chapter 9 and Chapter 8 make important recommendations about community-level resources that are required to sustain arts-based programs.

In addition to contributing to multi-dimensional conceptualizations of social inclusion, the chapters in this book draw on theories of human flourishing (Oberholzer, 2019; Seligman, 2011) which are seldom applied to later life (Kontos and Grigorovich, 2018; Momtaz et al., 2015). The concept of human flourishing seeks to capture an optimal state of well-being. In contrast to biomedical understandings of health, the absence of physical and mental illness is not a requirement for human flourishing. Scholars of philosophy, psychology, and other social sciences emphasize positive relationships (e.g., supportive relationships, opportunities to contribute to the happiness of others), engagement and interest in one's activities, meaning and purpose in life, and positive emotions (e.g., feeling respected) (Diener et al., 2010; Oberholzer, 2019; Seligman, 2011) as central elements of human flourishing. The relative lack of research on human flourishing in later life reflects deficit views of older adults which focus on loss rather than life enrichment and inclusion.

Arts-based ageing research and programming can enhance understandings and experiences of human flourishing in later life. As the chapters in this book show, dance can support meaningful self-expression, affect, and relationality (see Kontos and Grigorovich, Chapter 7). At a programmatic level, Rachel Bar and Pia Kontos in Chapter 4 explain that the traditional structure of a Sharing Dance class supports the expression of gratitude (e.g., révérénce) and contributes to the development of social relationships through technical aspects such as eye-line that encourage participants to make eye contact with one another. In Chapter 7, Pia Kontos and Alisa Grigorovich explain further that narrative approaches to dance can facilitate playfulness and sociability by encouraging dancers to use their imagination and participate in a shared story in their own creative way. Each of these aspects of Sharing Dance illustrates the role of arts-based programming in enhancing relationships and supporting engagement and meaning in later life. Moreover, rather than see the arts as a tool to address problems of ageing, human flourishing redirects attention to the broader benefits of arts-based initiatives for individuals.

Methodological contributions

To explore the many contributions of Sharing Dance the collaborative research profiled in this book involved multiple qualitative methods. Collecting multiple types of qualitative data (as well as involving multiple disciplines and community perspectives) allowed the research team to crystallize different facets of the Sharing Dance experience, program delivery, and challenges related to expanding the program to develop a more in-depth and complex understanding of the impact of the program. As noted above, the multi-method qualitative sequential research design contrasted with the more outcome-focused quantitative research that is dominant in much of the research on ageing, arts, and health (Parsons et al., 2017; Skinner et al., 2018). The team wanted to understand not just the outcomes of participating in Sharing Dance but also the processes involved in shaping those outcomes. In addition to being multi-method, the research was also characterized by multi-vocality, the inclusion of multiple and "varied voices" (Tracy, 2010, p. 844). Instead of the focus on the impact of Sharing Dance on a particular population and a particular behaviour or condition, the research explored the impact from the perspective of older adults themselves, family carers, volunteers, and organizations hosting the program to gain a more holistic and contextualized understanding of impact. By combining interviews, diaries, focus groups, and observations, as a team we were able to hear and see different experiences as well as challenges associated with Sharing Dance at multiple points and from multiple perspectives enriching understanding of the impact and supporting the continued development of Sharing Dance. The research in the book contributed to a smaller body of collaborative multi-method qualitative studies on arts-based programs (Kontos et al., 2020b). In addition, the book offered strategies and approaches for engaging in collaborative arts-based research by sharing recurrent themes from weekly researcher reflections on the processes and practices required to bring together diverse stakeholders and perspectives.

Transdisciplinary and creative collaboration

This project highlighted both the opportunities and challenges that come with transdisciplinary and collaborative research projects. The insights highlighted throughout this book, specifically by Mark Skinner and Rachel Bar in Chapter 3, collectively display the ways in which a transdisciplinary approach can inform a project's methods and methodology. The tools used to inform this study were shaped by the diverse disciplines involved, including non-academic expertise from artists and community stakeholders. The approach within which the project was framed and data was analyzed was also directly informed by the different disciplines contributing to the study and fostered opportunity for creative collaboration.

It is difficult to know how this project would have developed without a transdisciplinary approach, but it is possible to recognize the benefits to this approach that have been demonstrated from other similarly framed projects. In Grigorovich et al.' (2018) scoping review of transdisciplinary research in ageing, health/medicine, and technology, the approach enhanced the integration of knowledge across disciplines. In the specific example of Gutman et al. (2009), a transdisciplinary approach was thought to be instrumental in facilitating successful collaboration across disciplines. In the current project, the research team worked with artists and community stakeholders so that each stakeholder's needs and hopes were considered throughout the development of the project and creatively addressed. Had the research team not been as receptive to the design integrity of the dance program and the ways in which its developers envisioned its creation, reports on this project, including this book, would have no doubt missed key details of the innovation such as the importance of eye-contact and musical choices. Had the artists not been receptive to researchers' observations and community stakeholder feedback, the program may have developed without the ability to be adapted for the different settings and abilities the program is now able to reach (e.g., rural communities). Arts-based ageing research by definition includes the knowledge and work of different disciplines. This project therefore highlights what others have already seen by demonstrating – that a transdisciplinary approach to arts-based ageing research supports the creative integration of knowledge across different disciplines.

Sequential and iterative approach

In Chapter 3, Mark Skinner and Rachel Bar identified ways in which this project's sequential, phased approach (Skinner et al., 2018) provided opportunity for this research project's methods and the innovation to optimally develop. This approach was a product of necessity for this project, but it ultimately proved a valuable approach for both the research project and the innovation. While arts-based research has often adopted iterative approaches to research (e.g., Watson, 2020), the approach emerged from engineering and software development (Jacobs and Graham, 2016). Rather than develop a software innovation with only preliminary data, an iterative approach offers developers the ability to respond to change as an innovation is being trialed, and thus allows for faster learning and product development (Jacobs and Graham, 2016). Sharing Dance is not a software but given that the delivery model of the program studied in this project relied on technology, the benefits of an iterative approach are not surprisingly similar to those understood by software developers. Similarly, the sequential method applied to this research project allowed for ineffective tools to be replaced and focus to be refined to maximize the collection of useful data. Thus, especially when focusing on technology-supported arts-based ageing research, a sequential, iterative design may prove beneficial for both the research project and the development of the innovation being studied.

Applied contributions

Challenging stigma

Throughout this book, authors have highlighted the ways in which both Sharing Dance and this research project have challenged age and dementia-related stigma. In Chapter 4, Rachel Bar and Pia Kontos speak to the ways in which the innovation centres around the art form and not strictly its therapeutic benefits. The approach of the program itself challenges the belief that people living with dementia may only access the arts within a therapeutic context. The chapter also addresses the ways in which this project contrasts with research that often further perpetuates stigma, by focusing almost exclusively on health-related benefits of arts-based innovations for older adults and people living with dementia rather than the meaning associated with the process of creating and expressing oneself through art.

This research project was also able to capture some of the specific ways in which opportunities to dance can challenge individuals' beliefs about older adults and people living in dementia. Whitney Strachan's reflection in Chapter 6 perhaps provides some of the most tangible evidence of this finding. Doubt and scepticism were some of Whitney's initial reactions to dance for older adults with dementia, but this quickly changed with the experience of dancing with her clients. Through this process, she reflected on how dance helped her learn more about the abilities of people living with dementia. This experience challenges the dominant messages about living with dementia that are explored by Rachel Bar and Pia Kontos in Chapter 4, and Pia Kontos and Alisa Grigorovich in Chapter 7 and speaks to a unique contribution dance can make to challenging and changing beliefs about people living with dementia. It is our hope that by studying the experiences and processes of an art-centred innovation, this book shines light on approaches to arts-based ageing research that challenge diminishing beliefs about older adults and people living with dementia and how arts-based innovations may be studied.

Creativity-based opportunities

One way in which Sharing Dance facilitated accessible dance opportunities for older adults was by constructing a creativity-based offering. In Chapter 2 Rachel Bar and John Dalrymple, and Chapter 4, Rachel Bar and Pia Kontos explain how there is no right or wrong way to participate in Sharing Dance and Pia Kontos and Alisa Grigorovich in Chapter 7 captures participants experiences with this approach. While Canada's National Ballet School (NBS) may be best known for the "Big C" creativity of dancers in its professional ballet program, Sharing Dance presents older adults with the opportunity to enjoy and play with "little c" creativity. The creative and often narrative presentation of dance in Sharing Dance group classes also provided opportunity for co-creativity as Pia Kontos and Alisa Grigorovich note in Chapter 7, thus fostering opportunity for relational experiences. Through this approach, it becomes apparent that dance, when presented with a creative approach, supports key components of human flourishment like sense of accomplishment and positive emotions. Thus, through studying Sharing Dance, we see that a creative and narrative approach to the arts may be shared with older people in a way that supports the flourishment of all who participate.

Next steps

For future dance programming

Among the many learnings outlined in this book, this project captured the joy and beauty that can be shared when dance is made accessible to older adults, including people living with dementia. With this clear understanding of its potential, the task now is for optimal implementation of dance opportunities.

From a dance programming perspective, different training initiatives to support the ability of the onsite facilitator to optimize participants' experiences of Sharing Dance are currently being developed. This study demonstrates that the facilitator of a remotely delivered dance class does not need a dance background to support a dance opportunity, but that they play a critical role in connecting the content on the screen with the dancers in the room with them (see An Kosurko, Ilkka Arminen and Melisa Stevanovic, Chapter 8). As we navigate a post-COVID-19 world, other arts-based initiatives for older adults may also benefit from considering how training carers/support persons in the room may optimize the digital delivery of arts-based opportunities for older adults.

With the learnings from this project in-hand, the most complex and critical next step is wider implementation. If dance indeed challenges age- and dementia-related stigma and fosters social inclusion, how do we get more older adults dancing? With knowledge translation/implementation being the final step of a research project, it is too often an afterthought. Even if an arts-based initiative is able to sustain itself in the settings in which it was studied, implementing beyond these settings takes additional time and resources. Research has demonstrated how theoretically informed knowledge translation strategies may prove valuable at this stage. Bar (2020) adopted the integrated-Promoting Action on Research Implementation in Health Services (i-PARIHS) framework (Harvey and Kitson, 2015) to inform the dissemination of research about dance for people living with Parkinson's disease (PD). The framework's handbook guides the facilitator through innovation, recipient, and context considerations that are essential for a successful implementation of health research. For example, it is especially important to consider who the recipients of the knowledge will be for the current project given the diverse disciplines and stakeholders needed to successfully implement dance opportunities for older adults. Dissemination of knowledge must target each of these groups of recipients appropriately. For example, administration in the older adult sector, dance teachers, older adults, and formal and informal carers are some of the groups who should have access to the learnings from this project.

In terms of effective dissemination tools, Bar (2020) also demonstrated how the i-PARIHS framework may be enhanced by adopting art-based dissemination approaches. Arts-based knowledge dissemination of health research has demonstrated great potential as a tool that supports an understanding of knowledge that goes beyond strictly processing information on a cognitive level (e.g., Kontos and Naglie, 2007; Lapum et al., 2014). For example, the opportunity to dance along with people with Parkinson's provided experiential and embodied learning that appeared to influence healthcare professionals' intentions to recommend dance to people with PD and encourage the PD community to do more to engage with the innovation (Bar, 2020). Similar theoretical and dissemination strategies may therefore support the implementation of dance initiatives for older adults more broadly.

In tandem with implementation considerations, the economic, social, and place-based challenges of sustaining arts-based programs must also be considered. At a macro-level, recent scoping reviews of the role of the arts in improving health suggest the need to co-finance arts-based programs by arts, health, and social care budgets to support collaboration and sustainability of programs (Fancourt and Finn, 2019). However, many arts programs and many arts-based research projects are the products of time-limited grants. This presents economic challenges to sustaining effective arts-based programming that are borne by organizations and individuals. In the case of Sharing Dance, the program was free to participants and organizations during the duration of the study and organizations and individuals were encouraged to continue to participate for a small cost-recovery fee afterwards. NBS also worked to develop financial assistance to support small non-for-profit organizations to continue to host the program. Despite this, some organizations declined to participate in future terms because of a lack of social resources such as volunteers and staff. As both Chapter 8 (Kosurko, Arminen, and Stevanovic.) and Chapter 9 (Menec, Skinner, and Kosurko) in this book pointed out, the time and knowledge of staff and volunteers on the ground were critical to the success of the program. Having appropriate infrastructure and training to use the required technology was also critical to the continued success of the program from the perspective of organizations and facilitators. Indeed, another critical consideration in sustaining arts-based programs is sustainability for whom? As the case of Sharing Dance attests to, it takes many different stakeholder groups to support the success and efficacy of a program in the short and long term. The unique challenges and strengths of these groups must be considered to promote sustainability. Furthermore, differences in power among these groups and groups who were not effectively engaged by the program must also be considered.

For future research

A recognized limitation of the *Improving Social Inclusion for People with Dementia and Carers through Sharing Dance* project, and indeed this book, was the lack of partners and participants from intersectional groups including people who identify as racialized and/or Indigenous, as well as those people who identify as, lesbian, gay, bisexual, transgender, intersex, and queer (LGBTIQ). While a strength of this research was partnerships and participation of groups that are marginalized by geography, disability, and age, more work is needed to make dance innovations for older adults more inclusive for all older adults. Future collaborative arts-based research should partner with diverse community groups interested in the development of accessible and meaningful dance opportunities for older adults in their communities. Recognizing, supporting, and enriching understandings of community-based dance practices grounded in the knowledge of diverse cultural groups would not only provide more access to dance opportunities for older adults, but also contribute to dance, ageing, and equity, diversity, and inclusion theory more broadly.

Future research should also continue to follow the implementation of this project's findings; particularly that older people and their communities benefit from access to dance opportunities. The research project at the heart of this book is not the first to note the potential of dance to improve the lives of older people, but access to dance opportunities remains elusive for many older Canadians and older adults internationally. Research should continue to explore opportunities and barriers to the implementation of dance and other arts-based innovations for older people. As this book outlines, implementation of arts-based innovations is complex. Supportive policies and funding may be essential enablers of success, but critical as well are the community stakeholders, like administration and staff who work with older people in creating a culture of dance for all. The more we can understand how to make dance accessible and sustainable, the more communities will enjoy the inherent benefits of these opportunities to flourish.

Concluding comments

As a conclusion, this chapter summarizes the theoretical, methodological, and applied contributions of the collaborative arts-based research project, *Improving Social Inclusion for People with Dementia and Carers through Sharing Dance*, that brought together more than 20 researchers, artists, older adults, and leaders who contributed to the creation of this unique book.

Specifically in this chapter, we showed how a collaborative arts-based research project advanced integrated and multi-dimensional models of social inclusion; we illustrated the value of creative collaborative research for enriching program development and theoretical understanding of Sharing Dance; and we showed how sharing dance supports stigma reduction and human flourishing. Looking to the future, we recognize that sharing knowledge about the development and impacts of Sharing Dance is not enough. More work is needed to support a successful implementation of arts-based programs and extend access to marginalized groups who were not reached within this study. It is our hope that the insights and considerations throughout this chapter and this book provide those moving through collaborative arts-based ageing initiatives and research with greater insight into the complexity and beauty that awaits them.

References

Bar, R. J. (2020). *Dance for Parkinson's: A project exploring arts-based knowledge dissemination informed by the i-PARIHS framework.* [Doctoral dissertation]. Toronto Metropolitan University.

Camic, P. M., Tischler, V., and Pearman, C. H. (2014). Viewing and making art together: A multi-session art-gallery-based intervention for people with dementia and their carers. *Aging & Mental Health, 18*(2), 161–168. https://doi.org/10.1080/13607863.2013.818101

Chen, Y. R. R., and Schulz, P. J. (2016). The effect of information communication technology interventions on reducing social isolation in the elderly: A systematic review. *Journal of Medical Internet Research*, 18(1), e18. https://doi.org/10.2196/jmir.4596

Diener, E., Wirtz, D., Tov, W., Kim-Prieto, C., Choi, D., Oishi, S., and Biswas-Diener, R. (2010). New well-being measures: Short scales to assess flourishing and positive and negative feelings. *Social Indicators Research, 97*(2), 143–156. https://doi.org/10.1007/s11205-009-9493-y

Dow, B., Haralambous, B., Hempton, C., Hunt, S., and Calleja, D. (2011) Evaluation of Alzheimer's Australia Vic Memory Lane Cafes. *International Psychogeriatrics*, 23, 246–255.

Dupuis, S. L., Kontos, P., Mitchell, G., Jonas-Simpson, C., and Gray, J. (2016). Re-claiming citizenship through the arts. *Dementia, 15*(3), 358–380. https://doi.org/10.1177/1471301216637206

Fancourt, D., and Finn, S. (2019). *What is the evidence on the role of the arts in improving health and well-being? A scoping review.* World Health Organisation Europe, Health Evidence Synthesis Report, 67.

Grigorovich, A., Fang, M. L., Sixsmith, J., and Kontos, P. (2018). Defining and evaluating transdisciplinary research: Implications for aging and technology. *Disability and Rehabilitation: Assistive Technology*, 14(6). https://doi.org/10.1080/17483107.2018.1496361

Gutman, M. A., Barker, D. C., Samples-Smart, F., and Morley, C. (2009). Evaluation of active living research: Progress and lessons in building a new field. *American Journal of Preventative Medicine, 36*(2), S22–S33, e7. https://doi.org/10.1016/j.amepre.2008.10.009.

Harvey, G., and Kitson, A. (2015). *Implementing Evidence-Based Practice in Healthcare: A Facilitation Guide*. London: Routledge.

Herron, R. V., Kelly, C., and Aubrecht, K. (2021). A conversation about ageism: Time to deinstitutionalize long-term care? *University of Toronto Quarterly*, *90*(2), 183–206. https://doi.org/10.3138/utq.90.2.09

Jacobs, M. A., and Graham, A. L. (2016). Iterative development and evaluation methods of mHealth behavior change interventions. *Current Opinion in Psychology*, *9*, 33–37.

Kontos, P. C. (2012). Rethinking sociability in long-term care: An embodied dimension of selfhood. *Dementia*, 11(3), 329–346.

Kontos, P., and Grigorovich, A. (2018). Rethinking musicality in dementia as embodied and relational. *Journal of Aging Studies*, *45*, 39–48. https://doi.org/10.1016/j.jaging.2018.01.006

Kontos, P. C., and Naglie, G. (2007). Expressions of personhood in Alzheimer's disease: An evaluation of research-based theatre as a pedagogical tool. *Qualitative Health Research*, *17*(6), 799–811.

Kontos, P., Grigorovich, A., Dupuis, S., Jonas-Simpson, C., Mitchell, G., and Gray, J. (2020a). Raising the curtain on stigma associated with dementia: Fostering a new cultural imaginary for a more inclusive society. *Critical Public Health*, *30*(1), 91–102. https://doi.org/10.1080/09581596.2018.1508822

Kontos, P., Grigorovich, A., Kosurko, A., Bar, R. J., Herron, R. V., Menec, V. H., and Skinner, M. W. (2020b). Dancing with dementia: Exploring the embodied dimensions of creativity and social engagement. *The Gerontologist*, *61*(5), 714–723. https://doi.org/10.1093/geront/gnaa129

Kontos, P., Radnofsky, M. L., Fehr, P., Belleville, M. R., Bottenberg, F., Fridley, M., Massad, S., Grigorovich, A., Carson, J., Rogenski, K., Carpenter, K. S., Dupuis, S., Battalen, J., McDonagh, D., Fassbender, K., and Whitehouse, P. (2021) Separate and unequal: A time to reimagine dementia. *Journal of Alzheimer's Disease*, *80*(4), 1395–1399. https://doi.org/10.3233/jad-210057

Lapum, J. L., Liu, L., Church, K., Yau, T. M., Ruttonsha, P., Matthews David, A., and Retta, B. (2014). Arts-informed research dissemination in the health sciences: An evaluation of peoples' responses to "The 7,024 th Patient" art installation. *Sage Open*, *4*(1). https://doi.org/10.1177/2158244014524211

Momtaz, Y., Hamid, T. A., Haron, S. A., and Bagat, M. F. (2015). Flourishing in later life. *Archives of Gerontology and Geriatrics*, *63*, 85–91. https://doi.org/10.1016/j.archger.2015.11.001

Moody, E., and Phinney, A. (2012). A community-engaged art program for older people: Fostering social inclusion. *Canadian Journal on Aging*, 31(1), 55–64. https://doi.org/10.1017/S0714980811000596

Oberholzer, A. E. (2019). Human flourishing in the lives of children living with a disability. *Journal of Disability & Religion*, 23(2), 127–148.

Parsons, J., Gladstone, B., Gray, J., and Kontos, P. (2017). Re-conceptualizing 'impact' in art-based health research. *Journal of Applied Arts and Health*, *8*, 155–173. https://doi.org/10.1386/jaah.8.2.155_1

Pinkert, C., Köhler, K., Von Kutzleben, M., Hochgräber, I., Cavazzini, C., Völz, S., Palm, R., and Holle, B. (2019). Social inclusion of people with dementia – An integrative review of theoretical frameworks, methods and findings in empirical studies. *Ageing and Society*, 1–21. https://doi.org/10.1017/S0144686X19001338

Seligman, M. E. (2011). Flourish: A visionary new understanding of happiness and well-being. *Policy*, 27(3), 60–61.

Skinner, M. W., Herron, R. V., Bar, R. J., Kontos, P., and Menec, V. (2018). Improving social inclusion for people with dementia and carers through sharing dance: A qualitative sequential continuum of care pilot study protocol. *BMJ open*, *8*(11), e026912. https://doi.org/10.1136/bmjopen-2018-026912

Tracy, S. J. (2010). Qualitative quality: Eight "big-tent" criteria for excellent qualitative research. *Qualitative inquiry*, *16*(10), 837–851. https://doi.org/10.1177/1077800410383121

Walsh, K., O'Shea, E., and Scharf, T. (2019). Rural old-age social exclusion: A conceptual framework on mediators of exclusion across the lifecourse. *Ageing & Society*, *40*(1), 2311–2337. https://doi.org/10.1017/S0144686X19000606

Warburton, J., Cowan, S., and Bathgate, T. (2013). Building social capital among rural, older Australians through information and communication technologies: A review article. *Australasian Journal on Ageing*, *32*(1), 8–14. https://doi.org/10.1111/j.1741-6612.2012.00634.x

Watson, A. (2020). Methods braiding: A technique for arts-based and mixed-methods research. *Sociological Research Online*, *25*(1), 66–83. https://doi.org/10.1177/1360780419849437

Zeilig, H., Tischler, V., van der Byl Williams, M., West, J., and Strohmaier, S. (2019). Co-creativity, well-being and agency: A case study analysis of a co-creative arts group for people with dementia. *Journal of Aging Studies*, *49*, 16–24. https://doi.org/10.1016/j.jaging.2019.03.002.

Coda

Sharing Dance with older adults during COVID-19

Sheena Campbell, Cassy Borth, Jenn Kairies, and Jennifer Killing

Introduction

As a tribute to the significance of the arts-based collaboration introduced in this book, in this Coda, we share additional stories of Canada's National Ballet School's (NBS) Sharing Dance Older Adults program from our perspective as leaders in innovative long-term residential care (LTRC) practice at *peopleCare*, a family-owned group of LTRC and retirement homes in Ontario, Canada. Specifically, we present the story of our organization's involvement in a subsequent expansion of the Sharing Dance program undertaken during the COVID-19 pandemic. In doing so, we hope to highlight the continuing importance of arts-based programs, such as those developed by NBS and their partners, and collaborative arts-based ageing research, such as that undertaken by the co-editors and contributing authors of this book, for improving the well-being of older adults living in LTRC and other congregate care settings in Canada and internationally.

A new partnership opportunity

peopleCare Communities is a mission-driven, values-based, family-owned leading operator of communities for older adults, with a 50-year history of service and innovation in Ontario's publicly funded LTRC sector. We develop and operate campuses of care that enable older adults to age-in-place, with their care needs met over time in a range of housing options for older adults such as LTRC, independent or retirement living, and assisted care. More than building exceptional quality residences for older adults, our vision is to *Change the World of Senior Living*.

While partnerships have always been a peopleCare strength, our journey to support our residents, staff, families, and partners during the COVID-19 pandemic has taken us down some amazing paths. The particular story in this Coda begins in Summer 2020 when, amid Canada's first wave of the COVID-19 pandemic, we were connected to the Director of Research and Health at NBS (book co-editor, Rachel Bar) through an introduction from our mutual partner at Trent University (book co-editor, Mark Skinner) with whom peopleCare had already established an exciting partnership to develop

DOI: 10.4324/9781003260691-11

a new LTRC home as part of Trent's University-Integrated Seniors Village. The team at NBS told us about an incredible dance program they had developed in partnership with Baycrest, called Sharing Dance. They asked whether peopleCare would be interested in piloting a virtual delivery version of the program's *In Your Seat* "Dance at Home" version in our homes.

As a family-owned organization with a long history of engagement and collaboration to foster meaningful experiences for residents, families, staff, partners, volunteers, students, and our communities, the opportunity to team up with NBS aligned with our values as an organization. We were also eager to help expand the Sharing Dance program beyond the original parameters of its evolution at NBS (so eloquently explained by Rachel Bar and John Dalrymple in Chapter 2) and the Canadian Institutes of Health Research (CIHR)/Alzheimer Society of Canada *Improving Social Inclusion for People with Dementia and Carers through Sharing Dance* project that inspired this collaborative book, since bringing creativity, music, and the arts to our residents was in perfect alignment with peopleCare's ongoing commitment and goal during the COVID-19 pandemic to balance resident safety and care quality, without ever compromising the quality of life.

With the global pandemic declared on March 11, 2020, it was not long before statistics revealed the canary in the coal mine about the threat of the COVID-19 virus to LTRC residents, with hundreds of outbreaks and thousands of deaths. As a leading organization in care for older adults that cares for over 1,000 of Ontario's most vulnerable people in our LTRC and retirement homes annually, peopleCare quickly actioned a decisive response to protect our residents and frontline teams.

Early on in Canada's first wave of the COVID-19 pandemic, the Province of Ontario took the unprecedented step of restricting visitors – including families – to create an "iron ring" of protection around LTRC. For our residents, families are that essential extra set of loving hands, eyes, and ears every day – a true caregiving partner with our teams. Having to stay away from our homes was heartbreaking. It also added to the sense of loss and isolation faced by residents who were used to the typical hustle and bustle of lively homes and the opportunity to spend time with their families, students, volunteers, and friends.

peopleCare wasted no time reaching out to Family Councils Ontario (an Ontario Ministry of Long-term Care funded non-profit association that works with residents' family, Family Councils, and home staff across the province to improve quality of life in LTRC) to talk about what they were hearing from families, how we could better support residents and their loved ones at this strange and challenging time, and what it might take to safely bring families back in LTRC homes. We were determined to honour and respect the meaningful role of families as our caregiving partners and maintain their trust and confidence. Most importantly, in alignment with our commitment to a relational approach to care, we wanted to keep residents and their families connected and engaged to mitigate any impacts of isolation on their mental health and well-being.

Informed by a consultation survey with family members of peopleCare LTRC residents, our initial strategies included virtual visits – practically unheard of prior to the pandemic. We also leveraged and strengthened our existing Facebook community of families and staff, creating an online social media campaign called #HeartfeltMoments, to share and celebrate the ways our teams, families, friends, and neighbours were supporting one another and ensuring our homes remained places of joy and resilience.

We wanted to do even more.

The peopleCare team was very excited about piloting the expansion of Sharing Dance in our homes, and what it could add to our residents' lives at this difficult time. Even the language used to describe the program was aspirational, offering residents a chance to play and express themselves through creative movement and music. It certainly felt like serendipity. The exact right thing we needed, at the exact right time, to bring a little joy into our residents' lives. We could not wait.

The who and how of the peopleCare pilot

Matt Mullenweg, founder of WordPress, once famously said "Technology is best when it brings people together." Sharing Dance at Home virtual pilot was something else altogether: an opportunity to bring the transformative power of creative movement, music, and artistic expression to our residents, in a safe and accessible way, during a pandemic no less.

In accordance with peopleCare ethics approval and informed consent protocols, beginning July 2020, residents and staff in two peopleCare LTRC homes and one retirement home began testing out NBS' streamed version of Sharing Dance. The six-week "at-home" pilot was similar to the streamed group programming they had been running before the pandemic but was shorter, offered as weekly 20-minute seated dance classes, taught by two NBS on-screen dance instructors (OSI) and accompanied by a live musician. Each new lesson built on the previous week and offered a different range of motion options to support inclusion and accessibility.

The details of how the pilot was run – for example, group sizes and whether offered in resident rooms or shared activity spaces – were different in every peopleCare Home, in part due to constantly changing directives and Public Health guidance that differed between jurisdictions.

Since peopleCare families would normally be invited to participate in this type of activity, in-person with their loved one in our homes, the flexibility of the online classes and that they were open to all was a tremendous benefit. The details and links were shared with families, with some taking part at home in the early days just for the fun of it. Our program leaders also worked to support family members to participate in a joint session with their loved one virtually, for example, through apps such as Skype and FaceTime, although uptake on this option was limited.

peopleCare residents and staff in these initial homes worked with NBS and the evaluators they employed to evaluate this version of the program. The evaluation included both online surveys participants could fill out as well as phone interviews. Additionally, the evaluators ran a focus group with staff facilitators specifically from peopleCare as part of the evaluation. This informed improvements ahead of a broader roll-out of the program across peopleCare's seven LTRC and one retirement home.

To illuminate the importance and impact of the Sharing Dance pilot with peopleCare during COVID-19, in the section below, we describe the experiences of some of our team members and residents as they brought Sharing Dance into our LTRC homes in Summer and Fall 2020. With their informed consent, and in an effort to convey the authenticity of their experiences, we present the voices of our team members and residents' verbatim.

One home's experience

Cassy Borth is the Director of Programs at peopleCare's Oakcrossing Long-Term Care home in London, Ontario. She was an early adopter of the Sharing Dance pilot with peopleCare, participated in the evaluation phase, and continues to see great value in what the program brings to residents, staff, and families.

> During the 10 weeks of the pilot, we facilitated the dance program every Monday at 10 a.m. We had a group of about 15 men and women who looked forward to this weekly time. The videos worked really well for our residents. They are quick, only 20 minutes long, so whether someone has a cognitive impairment or dementia, the videos kept them completely engaged throughout. The music was also wonderful in keeping the residents engaged, it wasn't the same every week. Our residents truly loved it.

In the early days, when LTRC homes were closed to all visitors and families, the Oakcrossing team sent the link to families and friends of the residents several times, and more than a few family members enjoyed the shared virtual experience. Once the pilot was completed, Cassy ran the program intermittently for another nine months or so in groups of 5–15 people at a time, depending on the current pandemic restrictions.

Stories worth sharing

To fully understand the positive impact that Sharing Dance had in our homes, we asked peopleCare's recreationists to share some stories from their experience in piloting the program. These particular stories from three different peopleCare homes stood out.

"Irene" is in her late 70s and lives in a peopleCare LTRC home due to progressive dementia. *Enthusiastic about art, oil painting and nature, Irene wrote novels when she was younger and reading remains a favourite pastime. Restless, often walking for hours on end, the restrictions of COVID-19 and not being outside or seeing her family were particularly hard on Irene. One day during the*

Sharing Dance pilot, a recreation team member joined Irene on her walk, and accompanied her to the program area where a virtual dance class was in session. When Irene heard the music and saw the movement, she was instantly engaged. She loved the 50s and 60s beach party that was the day's theme, moving her arms and legs in time with the music. As the recreation staff describes it: "She was very involved, smiling, laughing and reminiscing. She walked away telling us how much she thoroughly enjoyed it. I'd say no doubt for her there was great value in this dance program."

We were a little surprised by the participation level of the men *– some originally would laugh and say "Heck no, I'm not dancing." Especially when they heard it was through a ballet school. We needed a different spin on things like telling them it was an exercise class with great music. Sometimes they came by just to see what the ladies were up to but then took part halfway through, once they saw what the program was like and that they would enjoy it. One of our gentlemen loves musical performances and singing and will come and listen to these types of programs but does not participate. He was 100% active during the dance program, tapping his toes, trying all the movements and really getting into the swing of things.*

In the height of the COVID-19 lockdown, we decided to organize a Zoom call and invite families to join us for our virtual dance classes. *We had no expectations, we simply hoped families would see this as an opportunity to engage in a joint online activity with their loved one. We were thrilled by the response, with four or five family members joining each time. It was really great because although our families were not able to be in person with us – accompanying their loved one to programs in the Home like they would in "normal times" – everyone made the best of it, participating in the class at the same time and really enjoying themselves. One of the family members said, "Thank you so much for making the pandemic feel as normal as it can be." Another emailed afterwards to say how appreciative she was to see her mom interacting and smiling with other residents. The residents themselves were very impressed with this technology and left the class grinning ear to ear.*

What we learned, what we shared

At the end of each of the two pilot phases, peopleCare worked with NBS to evaluate the expanded Sharing Dance program with the aim of making it more meaningful and impactful for participants.

The first opportunity to share peopleCare's experience of bringing Sharing Dance to our residents was through informal feedback provided by Jenn Kairies, former Director of Programs at peopleCare Tavistock Long-Term Care, who was the organization's lead and champion for this pilot. Below is a synopsis of what she shared with NBS on behalf of her peers:

> Our residents are absolutely loving the program. It's simple, enjoyable and IPAC (infection prevention and control) friendly. The electronic delivery format offers us maximum flexibility – big screen, small screen,

> it's easy to run in small groups or in an individual's room. Having the week's schedule posted for a good length of time helps us to run the classes when most convenient for residents. The only suggestion so far is that for our residents, many of whom have cognitive impairments, it's a bit difficult to follow along when there are two instructors, but otherwise, the program is great.

It should be noted that peopleCare's team members consistently found NBS to be extremely responsive and open to suggestions that would enhance the program for our residents. As a result of the feedback provided after the first phase of the pilot, NBS recorded new sessions for use in peopleCare homes that included only one instructor, doing one set of movements, accompanied by the pianist. This was much easier for residents to follow along with and supported greater engagement with the NBS on-screen instructor (OSI) (versus watching and mimicking their own, more familiar program staff in the Home).

After the second pilot phase, program directors and recreationists from four of peopleCare's eight homes – nine individuals in total from three LTRC and one retirement home – provided feedback during a video conference focus group. The purpose of the meeting was to understand the context in which different peopleCare staff and residents were using the Dance at Home videos and hear more about their experiences and the impact on various outcomes such as physical health, mood, and social connections. The focus group also served to help NBS further understand successes and challenges related to using the videos specifically in these settings.

NBS has commented that it was particularly touching to see one peopleCare team show up to the focus group in full personal protective equipment (PPE) – gowns, gloves, masks, and face shields – as they were battling a significant COVID-19 outbreak in the home. This was typical of peopleCare's dedicated frontline – stepping up with determination not to compromise our focus on changing care for older adults in LTRC or let COVID-19 get the better of them.

The focus group formed a small component of NBS' larger formal evaluation and report. Once again, NBS adapted the program based on feedback about accessibility considerations (i.e., different leg movements for participants in wheelchairs with foot pedals and a standing option to make the program more accessible to diverse bodies). peopleCare staff provided additional comments about the program saying:

> The best part for our residents is that there was no right or wrong way to do it. And even a little participation goes a long way – they could get those arms and legs moving fully. Or they could sit and listen to and feel the music. Not to mention how it motivates residents. Unlike many physical activities that can tire them out quickly, residents were happy to keep going a bit longer because of the music. The sense of playfulness and fun were hugely important and made all the difference in keeping participants engaged.

A definite advantage of the Sharing Dance at Home program is the virtual aspect. Pre-recorded and posted on the internet for on-demand streaming or downloading, the flexibility to deliver the program using a computer, phone, or tablet, at any time, to any number of residents, in any size or configured setting, was a tremendous benefit to peopleCare's programs teams, who were determinedly leading through the uncertainty of the COVID-19 pandemic with creativity and a can-do spirit.

For the purposes of the pilot, peopleCare focused on supporting, enhancing, and observing the experiences of our residents, staff, and families through their participation in the program.

The value of artistic expression and creative movement to our residents

A big part of what makes peopleCare homes special is a focus on programs that offer social interaction, meaningful time and new relationships across the generations. A great example is our partnership with Western University to co-house music students in our retirement home. As Betty Anne Younker, past Dean of the Don Wright Faculty of Music, said about our partnership: "Stories are shared, life lessons offered and insights provided. The residents enjoy the vibrancy of the youth, the stories they tell and of course, the music that fills the spaces in which they gather."

An important benefit peopleCare saw from the Sharing Dance program is how it supported staff and resident engagement (an intergenerational interaction in itself). It allowed residents and staff to enjoy participating together versus the staff performing the program "or" the residents. As conveyed by Jenn Kairies, "Many laughs and jokes were shared as both the residents and the staff became more confident in the program and grew accustomed to the movements." Indeed, according to Jenn Kairies, from a recreation therapist's point of view, Sharing Dance offered multiple benefits consistent with holistic approaches to therapeutic recreation that support human flourishing (Anderson and Heyne, 2012). For instance, physically the pilot enabled safe and accessible creative movement, while supporting health and well-being. Emotionally, the pilot enabled residents to comfortably express thoughts and feelings in an inclusive atmosphere, promoting healthy expression of emotions and feelings of achievement, and building a sense of community. Cognitively, the pilot enabled residents to learn and function intellectually; it provided residents with the opportunity to practise decision-making and strategic thinking while implementing movements, and residents were able to watch, learn, and mimic the dancers. Socially the pilot enabled residents to enjoy meaningful relationships by providing an opportunity to engage socially with others, and it allowed sharing of thoughts and opinions. Finally, spiritually, the pilot enabled residents to find purpose in life, fostered creative expressions through purposeful movement, served as an inspiration to residents, and was offered in a non-threatening environment to promote self-expression.

Remembering the good times

For older adults, reflecting on the past in a positive way can help maintain a sense of identity and self. Across our homes, staff commented on how strongly Sharing Dance supported reminiscing, and the boost in mind and spirit this gave to our residents. The music brought them back to another time in their lives, often a time of love and connectedness.

At a pilot session in one of our homes, a dozen plus residents joined in the fun of scooping ice cream, as the onscreen dance instructor encouraged them to recall childhood days of summer and their favourite flavour melting on their tongues. Music drifted down the hall where others sat tapping their toes with smiles on their faces.

Asked her thoughts afterwards, one participant enthused about the memories the music invoked of many former happy times dancing with a dear friend: "It brought tears to my eyes – that song was one that my friend and I would dance to all the time."

Residents living with dementia

According to the Ontario Long-Term Care Association, in their 2019 This is Long-Term Care Report, the proportion of LTRC residents living with dementia has been growing steadily since 2010, with two out of every three residents (64%) now affected by these diseases. And in fact, 90% of all LTRC residents have some form of cognitive impairment. Delivering resident-centred emotion-based care that connects to the person in an individualized and dignified way, supports the quality of life for each person living with dementia.

Many individuals living with dementia respond emotionally and physically to stressors in their environment with actions such as wandering, agitation, and anger. In several of our homes, the Sharing Dance program was seen as particularly beneficial to residents living with dementia. The simple movements were easy to follow, and the music was enjoyed by most. One Home noted they offered the program at various times in the day and found it assisted the residents in a soothing way, at times when their needs were higher:

> The purposeful movement seemed to lessen their need to wander all the time. We would try it out at different times a day, at night it was calming. One resident I would bring when I was doing the program was quite agitated but would be calmer afterward.

Changing care for older adults

When our residents speak positively about how it feels to experience life in a peopleCare Home, we know we are doing a good job. It is one of our most meaningful measures of success. During such a tough time – with isolation, visitor restrictions, and the need for masks and social distancing – bringing Sharing Dance into our homes was a direct reflection of the determination of our frontline teams to keep as much fun, love, and laughter in our homes as possible: "Sharing Dance gave our residents something to socialize about. They were still talking about the session days after. And they get excited mid-week because the dance program is coming up again."

During the COVID-19 pandemic, protective measures impacted residents' emotional health and well-being, requiring a concerted effort on the part of LTRC teams to draw on their wellspring of creativity and resiliency to ensure safety while supporting quality of life. As the pandemic has progressed, measures have evolved. For example, the uninterrupted presence of family carers is now enshrined as a resident right under the legislation.

COVID-19 is expected to become endemic – with some variants circulating in society year-round. Pre-pandemic, outbreaks of various illnesses impacted homes, closing admissions, and requiring targeted infection prevention and control measures, including isolation of some residents. Regardless of the type of illness causing an outbreak, necessary enhanced measures will always run the risk of impacting resident quality of life for the duration. LTRC operators will need to continue to find innovative ways to keep residents active and engaged regardless of circumstances.

One of the greatest advantages of Sharing Dance at Home is the absolute ease with which it can be adapted and applied in terms of settings, times, group sizes, participants and more – all to better suit the unique needs of each LTRC resident or groups of residents.

Any time restrictions hamper our ability to provide recreation and activation, our homes can find opportunities to engage residents through Sharing Dance – in their rooms, in small groups in the activity spaces, with one staff supporting, with their loved one present or joining in on the screen. No matter where or when, the program's music and dance have the power to transport anyone, anywhere – to a feeling, a time, or a place they want to be. The possibilities are endless.

Concluding comments

peopleCare's vision is to *Change the World of Senior Living* – in part by doing all we can to make the last chapter of our residents' lives more engaging and rewarding. We appreciate NBS and their partners, including those who are part of this book, for supporting our dedicated frontline teams to fulfil their purpose and to bring the pleasure and benefits of creative movement into our residents' lives at a time when it mattered so much. It was peopleCare's immense pleasure to contribute our perspective of Sharing Dance during COVID-19 to this important book by sharing the experiences of our leaders, staff and residents in this Coda, in hopes that it encourages and inspires others to consider the many positive impacts of incorporating arts-based programs, and dance in particular, as well as collaborative research into their LTRC homes to bring more joy into their residents' lives.

References

Anderson, L. S., and Heyne, L. A. (2012). Flourishing through leisure: An ecological extension of the leisure and well-being model in therapeutic recreation strengths-based practice. *Therapeutic Recreation Journal*, *46*(2), 129.

Ontario Long-Term Care Association. (2019). *This is Long-Term Care 2019*. https://www.oltca.com/OLTCA/Documents/Reports/TILTC2019web.pdf

Index

Note: Pages in *italics* refer figures and pages in **bold** refer tables.